Lectures in Mathematics
ETH Zürich
Department of Mathematics
Research Institute of Mathematics

Managing Editor:
Michael Struwe

Mark Jerrum

Counting, Sampling and Integrating: Algorithms and Complexity

Birkhäuser Verlag
Basel · Boston · Berlin

Author's address:

Division of Informatics
The University of Edinburgh
James Clerk Maxwell Building
King's Buildings
Edinburgh EH9 3JZ
Scotland, UK
E-mail: mrj@dcs.ed.ac.uk

2000 Mathematical Subject Classification 05Cxx, 60Jxx, 68Q17, 68Q25, 82B80

A CIP catalogue record for this book is available from the
Library of Congress, Washington D.C., USA

Bibliographic information published by Die Deutsche Bibliothek
Die Deutsche Bibliothek lists this publication in the Deutsche Nationalbibliografie; detailed bibliographic data is available in the Internet at <http://dnb.ddb.de>.

ISBN 3-7643-6946-9 Birkhäuser Verlag, Basel – Boston – Berlin

Member of the BertelsmannSpringer Publishing Group
Printed on acid-free paper produced from chlorine-free pulp. TCF ∞
Printed in Germany
ISBN 3-7643-6946-9

9 8 7 6 5 4 3 2 1 www.birkhauser.ch

In memory of my father.

Contents

Foreword

These notes had their origin in a postgraduate lecture series I gave at the Eidgenössiche Technische Hochschule (ETH) in Zürich in the Spring of 2000. I am very grateful to my hosts, the Forschungsinstitut für Mathematik at ETH, for providing the ideal opportunity to develop and present this material in what I hope is a reasonably coherent manner, and also for encouraging and assisting me to record the proceedings in these lecture notes.

The subject of the lecture series was counting (of combinatorial structures) and related topics, viewed from a computational perspective. As we shall see, "related topics" include sampling combinatorial structures (being computationally equivalent to approximate counting via efficient reductions), evaluating partition functions (being weighted counting) and calculating the volume of bodies (being counting in the limit).

We shall be inhabiting a different world to the one conjured up by books with titles like *Combinatorial Enumeration* or *Graphical Enumeration.* There, the problems are usually parameterised on a single integer parameter n, and the required solutions are closed form or asymptotic estimates obtained using very refined and precise analytical tools. An example problem might be "how many distinct unlabelled graphs are there on n vertices?"

Instead, we shall be working in a somewhat messier world of problems, ultimately inspired by practical applications, in which an instance (say, an undirected graph) is presented and one is asked to count certain structures (say perfect matchings) within it. (The "practical application" in this instance might be in the domain of statistical physics, where perfect matchings are configurations of the "dimer model".) A solution in this world is an efficient algorithm which takes as input the problem instance, and outputs the number of structures. As a first cut, "efficient" is taken to mean "having a running time that scales as a polynomial in some natural measure of the instance size". Later, particularly if the problem seems reasonably close to practical application, one might well be concerned to reduce the degree of the polynomial as far as possible.

These notes open with a couple of examples of combinatorial structures that can be counted exactly polynomial time. Although the examples are classical, I decided to include them because (i) they perfectly demonstrate the ideal we are aiming for, and (ii) the ideas on which they are based are beautiful and worth reflecting on. It transpires that these two examples are the exceptions that prove the empirical rule that exact counting is hard. The theory of #P-completeness, which we review briefly, provides evidence for the latter assertion. To make progress, we lower our sights by seeking solutions (i.e., algorithms) that are approximate only, but which nevertheless come with rigorously derived guarantees on the relative error of the results they produce. The notion of efficient approximation algorithm is formalised in the definition of "FPRAS".

Of the various approaches to designing efficient approximation algorithms for counting problems, by far the most fruitful has been "Markov chain Monte Carlo" (MCMC). The idea is to accumulate information on a set of combinatorial structures by performing a random walk (i.e., simulating a Markov chain) on those structures. The running time of an MCMC algorithm depends on the rate of convergence to equilibrium of this Markov chain, as formalised in the notion of "mixing time" of the Markov chain. To my mind, the most exciting developments in the area in recent years have been the introduction and refinement of various analytical tools for bounding the mixing time of combinatorially-defined Markov chains. A survey of these tools forms the primary theme of these notes. Geometric, canonical path and coupling arguments are all covered, together with illustrating examples.

Not every counting problem can be solved efficiently, even in the approximate FPRAS sense, and the notes close with some remarks on intractable problems.

Despite the relative youthfulness of the topic, any treatment of this length must necessarily miss out more than it includes. In particular, with the intention of getting close to the coal-face of current research, I have concentrated on the MCMC approach to the exclusion of others. Among the sacrifices this decision entailed is Karp and Luby's proposal for estimating the size of a union of sets [40]. Moreover, I have said nothing at all about the sampling and counting of unlabelled structures, despite a reasonably extensive, if somewhat provisional body of work on the subject (see surveys by Jerrum [34] and Goldberg [31]). Even within the MCMC theme my coverage is certainly not exhaustive. Among several authors who have a right to be offended by the omission of their work are Feder and Mihail [27], with their surprising and elegent inductive approach to counting and sampling matroid bases. The comparison method of Diaconis and Saloff-Coste [19] also escapes mention, as does the decomposition approach of Madras and Randall [48].

So much for intention; now for the means. It is a pleasure to record my gratitude to various research students and associates at the Institut für Theoretische Informatik at ETH for capturing so well the content of the lectures in written form, and ultimately as TEX source files. They are: Christoph Ambühl, Alex Below, Bernd Gärtner, Michael Hoffmann, Zsuzsanna Lipták, Samuele Pedroni and Uli Wagner. These files, together with fragments of earlier surveys, formed the raw material for this volume. I also gratefully acknowledge the contribution of Alistair Sinclair, who collaborated on a significant proportion of the work described in these pages.

But as well as thanking the note takers, I must also apologise to them. For the raw material they provided has undergone a series of reworkings, accumulating layers of extra material until an archaeologist would be required to uncover the ur-text. First, I realised that extra explanation would be required to make the material comprehensible in the absence of the spoken and somewhat interactive presentations. Then there was the matter of imposing some overall consistency of notation and terminology where it had been lacking in the lectures, and attempting some limited moves in the direction of a uniform style. Then there were the things I wish I had done slightly differently, and where the temptation to rewrite history was simply too great. Finally, I abandoned any attempt to mirror the division of the course into lectures in the structure of this volume. Anyway, the fact is that without their efforts there would be no book (and certainly no fancy illustrations to enliven the text).

Mark Jerrum, Edinburgh, April 2002.

Chapter 1

Two good counting algorithms

Counting problems that can be solved exactly in polynomial time are few and far between. Here are two classical examples whose solution makes elegant use of linear algebra. Both algorithms predate the now commonplace distinction between polynomial and exponential time, which is often credited (with justification) to Edmonds in the mid 1960s; indeed our first example dates back over 150 years!

1.1 Spanning trees

Basic graph-theoretic terminology will be assumed. Let $G = (V, E)$ be a finite undirected graph with vertex set V and edge set E. For convenience we identify the vertex set V with the first n natural numbers $[n] = \{0, 1, \dots, n-1\}$. The *adjacency matrix* A *of* G is the $n \times n$ symmetric matrix whose ij'th entry is 1 if $\{i, j\} \in E$, and 0 otherwise. Assume G is connected. A *spanning tree* in G is a maximum (edge) cardinality cycle-free subgraph (equivalently, a minimum cardinality connected subgraph that includes all vertices). Any spanning tree has $n - 1$ edges.

Theorem 1.1 (Kirchhoff). *Let $G = (V, E)$ be a connected, loop-free, undirected graph on n vertices, A its adjacency matrix and $D = \operatorname{diag}(d_0, \dots, d_{n-1})$ the diagonal matrix with the degrees of the vertices of G in its main diagonal. Then, for any i, $0 \le i \le n-1$,*

$$\#\text{ spanning trees of } G = \det(D - A)_{ii},$$

where $(D - A)_{ii}$ is the $(n-1) \times (n-1)$ principal submatrix of $D - A$ resulting from deleting the i'th row and i'th column.

Since the determinant of a matrix may be be computed in time $O(n^3)$ by Gaussian elimination, Theorem 1.1 immediately implies a polynomial-time algorithm for counting spanning trees in an undirected graph.

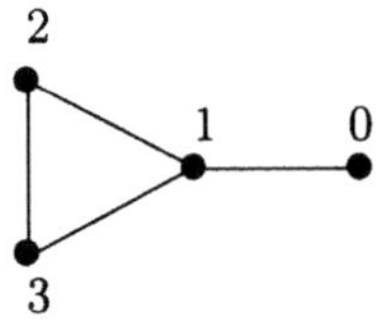

Graph G

$$\begin{pmatrix} 1 & -1 & 0 & 0 \\ -1 & 3 & -1 & -1 \\ 0 & -1 & 2 & -1 \\ 0 & -1 & -1 & 2 \end{pmatrix}$$

$D - A$

$$\begin{pmatrix} 1 & 0 & 0 \\ 0 & 2 & -1 \\ 0 & -1 & 2 \end{pmatrix}$$

$(D - A)_{11}$

Figure 1.1: Example illustrating Theorem 1.1.

Example 1.2. Figure 1.1 shows a graph G with its associated "Laplacian" $D - A$ and principal minor $(D - A)_{11}$. Note that $\det(D - A)_{11} = 3$ in agreement with Theorem 1.1.

Remark 1.3. The theorem holds for unconnected graphs G, as well, because then the matrix $D - A$ associated with G is singular. To see this, observe that the rows and columns of a connected graph add up to 0 and, similarly, those of any submatrix corresponding to a connected component add up to 0. Now choose vertex i and a connected component C such that $i \notin C$. Then, the columns of $(D - A)_{ii}$ that correspond to C are linearly dependent, and $(D - A)_{ii}$ is singular.

Our proof of Theorem 1.1 follows closely the treatment of van Lint and Wilson [65], and relies on the following expansion for the determinant, the proof of which is deferred.

Lemma 1.4 (Binet-Cauchy). *Let A be an $(r \times m)$- and B an $(m \times r)$-matrix. Then*

$$\det AB = \sum_{\substack{S \subseteq [m], \\ |S| = r}} \det A_{*S} \det B_{S*},$$

where A_{*S} is the square submatrix of A resulting from deleting all columns of A whose index is not in S, while, similarly, B_{S*} is the square submatrix of B resulting from B by deleting those rows not in S.

Remark 1.5. Typically, r is smaller than m. However, the lemma is also true for $r > m$. Then the sum on the right is empty and thus 0. But also AB is singular, since $\operatorname{rank} AB \leq \operatorname{rank} A \leq m < r$.

Let H be a directed graph on n vertices with m edges. Then the *incidence matrix of H* is the $(n \times m)$-matrix $N = (\nu_{ve})$ where

$$\nu_{ve} = \begin{cases} +1, & \text{if vertex } v \text{ is the head of edge } e; \\ -1, & \text{if } v \text{ is the tail of } e; \\ 0, & \text{otherwise.} \end{cases}$$

The *weakly connected components of H* are the connected components of the underlying undirected graph, i.e., the graph obtained from H by ignoring the orientations of edges.

Fact 1.6.

$$\operatorname{rank} N = |V(H)| - |\mathcal{C}(H)| = n - |\mathcal{C}(H)|,$$

where $V(H)$ *is the vertex set of* H *and* $\mathcal{C}(H) \subseteq 2^{V(H)}$ *is the set of (weakly) connected components of* H.

Proof. Consider the linear map represented by $N^\top$, the transpose of N. It is easy to see that, if h is a vector of length n, then

$$N^\top h = 0 \Leftrightarrow h \text{ is constant on connected components},$$

i.e., $i, j \in C \Rightarrow h_i = h_j$, for all $C \in \mathcal{C}(H)$. This implies that $\dim\ker N^\top = |\mathcal{C}(H)|$, proving the claim, since $\operatorname{rank} N = \operatorname{rank} N^\top = n - \dim\ker N^\top$. □

Fact 1.7. *Let* B *be a square matrix with entries in* $\{-1, 0, +1\}$ *such that in each column there is at most one* $+1$ *and at most one* -1. *Then,* $\det B \in \{-1, 0, +1\}$.

Proof. We use induction on the size n of B. For $n = 1$, the claim in trivial. Let $n > 1$. If B has a column which equals 0, or if each column has exactly one $+1$ and one -1, then B is singular. Otherwise there is a column j with either one $+1$ or one -1, say in its i'th entry b_{ij}, and the rest 0's. Developing $\det B$ by this entry yields $\det B = \pm b_{ij} \det B_{ij}$, where B_{ij} is the minor of B obtained by deleting row i and column j. By the induction hypothesis, the latter expression equals -1, 0 or $+1$. □

The ingredients for the proof of Kirchhoff's result are now in place.

Proof of Theorem 1.1. Let $\vec{G}$ be an arbitrary orientation of G, N its incidence matrix, and $S \subseteq E$ be a set of edges of $\vec{G}$ with $|S| = n - 1$. Then, by Fact 1.6,

$$\operatorname{rank}(N_{*S}) = n - 1 \Leftrightarrow S \text{ is the edge set of a tree.} \tag{1.1}$$

(The condition that S is the edge set of a tree again ignores the orientation of edges in S.) If N' results from N by deleting one row, then

$$\operatorname{rank}(N'_{*S}) = \operatorname{rank}(N_{*S}). \tag{1.2}$$

This is because the deleted row is a linear combination of the others, since the rows of N add up to 0. Combining (1.1) and (1.2) with Fact 1.7 gives us

$$\det N'_{*S} = \begin{cases} \pm 1, & \text{if } S \text{ is a spanning tree;} \\ 0, & \text{otherwise.} \end{cases} \tag{1.3}$$

Now observe that $D - A = NN^\top$, since

$$(NN^\top)_{ij} = \sum_{e \in E} \nu_{ie}\nu_{je} = \begin{cases} -1, & \text{if } \{i, j\} \in E; \\ d_i, & \text{if } i = j; \\ 0, & \text{otherwise.} \end{cases}$$

Clearly, $(D-A)_{ii} = N'(N')^\top$ where N' results from N by deleting any row i. Thus,

$$\begin{aligned}\det(D-A)_{ii} &= \det(N'(N')^\top)\\ &= \sum_{|S|=n-1} \det N'_{*S} \det((N')^\top)_{S*} && \text{by Lemma 1.4}\\ &= \sum_{|S|=n-1} \det N'_{*S} \det(N'_{*S})^\top\\ &= \#\text{ spanning trees of } G && \text{by (1.3).}\end{aligned}$$

□

It only remains to prove the key lemma on expanding determinants.

Proof of Lemma 1.4. We prove a more general claim, namely

$$\det A\Delta B = \sum_{\substack{S\subseteq[m],\\ |S|=r}} \det A_{*S} \det B_{S*} \prod_{i\in S} e_i,$$

where $\Delta = \operatorname{diag}(e_0,\dots,e_{m-1})$. The lemma follows by setting all e_i to 1. Observe that entries of $A\Delta B$ are linear forms in $e_0,\dots,e_{m-1}$. Thus, $\det A\Delta B$ is a homogeneous polynomial of degree r in $e_0,\dots,e_{m-1}$, i.e., all monomials have degree r. Comparing coefficients will yield the desired result. First we observe that every monomial in $\det A\Delta B$ must have r distinct variables. For if not, consider a monomial with the fewest number of distinct variables, and suppose this number is less than r. Setting all other variables to 0 will result in $\det A\Delta B = 0$, since $\operatorname{rank} A\Delta B \le \operatorname{rank} \Delta < r$ and $A\Delta B$ is singular. But $\det A\Delta B = 0$ implies that the coefficient of the monomial is 0. Now look at a monomial with exactly r distinct variables, say $\prod_{i\in S} e_i$. Set these variables to 1 and all others to 0. Then, $A\Delta B$ evaluates to $A_{*S}B_{S*}$, and hence the coefficient of $\prod_{i\in S} e_i$ is $\det A_{*S}B_{S*} = \det A_{*S} \det B_{S*}$. □

It is possible to generalise Theorem 1.1 to directed graphs $G = (V, E)$, where a directed spanning tree (or *arborescence*) is understood to be a subgraph $(V, T\subseteq E)$ where (i) (V, T) with the orientation of edges ignored forms a spanning tree of the unoriented version of G, and (ii) the orientations of edges in T are consistently directed towards some distinguished vertex or *root* r. Equivalently, it is an acyclic subgraph in which every vertex other than the distinguished root r has outdegree 1, and the root itself has outdegree 0. (There does not seem to be agreement on whether edges should be directed towards or away from the root; towards seems more natural — corresponding as it does to functions on $[n]$ with a unique fixed point — and in any case better suits our immediate purpose.)

An *Eulerian circuit* in a directed graph G is a closed path (i.e., one that returns to its starting point) that traverses every edge of G exactly once, respecting the orientation of edges. (The path will not in general be simple, that is to say it will visit vertices more than once.) The number of Eulerian circuits in a directed graph is related in a simple way to the number of arborescences, so these structures also can be counted in polynomial time. For details see Tutte [60, §VI.3, §VI.4].

Open Problem. To the best of my knowledge, it is not known whether there exists a polynomial-time algorithm for counting Eulerian circuits in an undirected graph. Note that the usual strategy of viewing an undirected graph as a directed graph with paired anti-parallel edges does not work here.

Exercise 1.8. Exhibit an explicit (constant) many-one relation between the Eulerian circuits in a directed graph G and the arborescences in G. Hint: use the arborescence to define an "escape route" or "edge of final exit" from each vertex.

1.2 Perfect matchings in a planar graph

Let $G = (V, E)$ be an undirected graph on n vertices ($V = [n]$, for convenience). A *matching* in G is a subset $M \subseteq E$ of pairwise vertex-disjoint edges. A matching M is called *perfect* if it covers V, i.e., $\bigcup M = V$. Note that n must be even for a perfect matching to exist.

Around 1960, Kasteleyn discovered a beautiful method for counting perfect matchings in a certain class of "Pfaffian orientable" graphs, which includes all planar graphs as a strict subclass. Linear algebra is again the key.

Fact 1.9. *If M, M' are two perfect matchings in G, then $M \cup M'$ is a collection of single edges and even (i.e., even length) cycles.*

Let $G = (V, E)$ be an undirected graph, C an even cycle in G, and $\vec{G}$ an orientation of G. We say that C is *oddly oriented by* $\vec{G}$ if, when traversing C in either direction, the number of co-oriented edges (i.e., edges whose orientation in $\vec{G}$ and in the traversal is the same) is odd. (Observe that the direction in which we choose to traverse C is not significant, since the parity in the other direction is the same.) An orientation $\vec{G}$ of G is *Pfaffian* (also called *admissible*) if the the following condition holds: for any two perfect matchings M, M' in G, every cycle in $M \cup M'$ is oddly oriented by $\vec{G}$. Note that all cycles in $M \cup M'$ are even.

Remark 1.10. The definition of Pfaffian orientation given above is not equivalent to requiring that all even cycles in G be oddly oriented by $\vec{G}$, since there may be even cycles that cannot be obtained as the union of two perfect matchings.

Let $\vec{G}$ be any orientation of G. Define the *skew adjacency matrix* $A_s(\vec{G}) = (a_{ij} : 0 \le i, j \le n-1)$ *of* G by

$$a_{ij} = \begin{cases} +1, & \text{if } (i,j) \in E(\vec{G}); \\ -1, & \text{if } (j,i) \in E(\vec{G}); \\ 0, & \text{otherwise.} \end{cases}$$

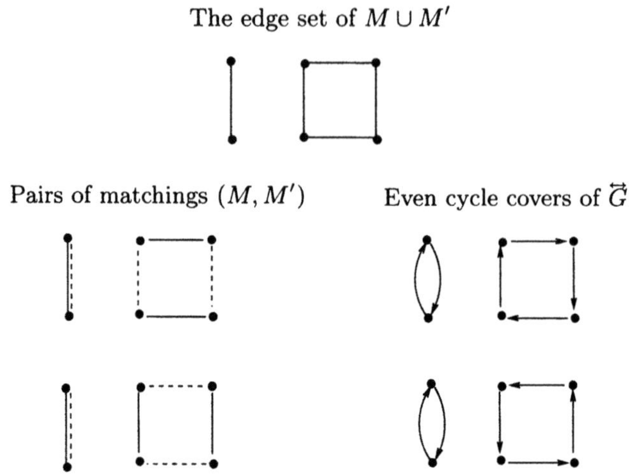

Figure 1.2: Bijection between pairs of matchings in G and even cycle covers of $\overleftrightarrow{G}$.

Theorem 1.11 (Kasteleyn). *For any Pfaffian orientation $\vec{G}$ of G,*

$$\#\text{ perfect matchings in } G = \sqrt{\det A_s(\vec{G})}\,.$$

Our proof of Theorem 1.11 borrows from Kasteleyn [41] and Lovász and Plummer [45]. Denote by $\overleftrightarrow{G}$ the directed graph obtained from G by replacing each undirected edge $\{i, j\}$ by the anti-parallel pair of directed edges $(i, j), (j, i)$. An *even cycle cover* of $\overleftrightarrow{G}$ is a collection $\mathcal{C}$ of even directed cycles $C \subseteq E(\overleftrightarrow{G})$ such that every vertex of G is contained in exactly one cycle in $\mathcal{C}$.

Lemma 1.12. *There is a bijection between (ordered) pairs of perfect matchings in G and even cycle covers in $\overleftrightarrow{G}$.*

Proof. Let (M, M') be a pair of perfect matchings in G. For each edge in $M \cap M'$ (i.e, each edge in $M \cup M'$ that does not lie in an even cycle) take both directed edges in $\overleftrightarrow{G}$. Now orient each cycle C in $M \cup M'$ (with length ≥ 4) according to some convention fixed in advance. For example, take the vertex with lowest number in C and orient the incident M-edge away from it. The resulting collection $\mathcal{C}$ of directed cycles is an even cycle cover of $\overleftrightarrow{G}$.

The procedure may be reversed. First, each oriented 2-cycle in $\mathcal{C}$ must correspond to an edge that is in both M and M'. Then, each even cycle $C \in \mathcal{C}$ of length at least four may be decomposed into alternating M-edges and M'-edges; the convention used to determine the orientation of C will indicate which of the two possible decompositions is the correct one. □

Proof of Theorem 1.11. In view of the previous lemma, we just need to show that $\det A_s(\vec{G})$ counts even cycle covers in $\overleftrightarrow{G}$. Now,

$$\det A_s(\vec{G}) := \sum_{\pi \in S_n} \operatorname{sgn} \pi \prod_{i=0}^{n-1} a_{i,\pi(i)}, \tag{1.4}$$

where S_n is the set of all permutations of $[n]$, and $\operatorname{sgn} \pi$ is the sign of permutation π.[1] Consider a permutation π and its (unique) decomposition into disjoint cycles $\pi = \gamma_1 \cdots \gamma_k$. Each γ_j acts on a certain subset $V_j \subseteq V$. The corresponding product $\prod_{i \in V_j} a_{i,\pi(i)}$ is non-zero if and only if the edges $\{(i, \pi(i)) : i \in V_j\}$ form a directed cycle in G, since otherwise one of the $a_{i,\pi(i)}$ would be 0. Thus, there is a one-to-one correspondence between permutations π with non-zero (i.e., ± 1) contributions to (1.4) and cycle covers in $\overleftrightarrow{G}$.

We now claim that sum (1.4) is unchanged if we restrict it to permutations with only even length cycles. To see this, consider a permutation π and an odd length cycle γ_j in π, say the first in some natural ordering on cycles. Let $\pi' = \gamma_1 \cdots (\gamma_j)^{-1} \cdots \gamma_k$ be identical to π except that γ_j is reversed. Then, $\prod_{i=0}^{n-1} a_{i,\pi(i)} = -\prod_{i=0}^{n-1} a_{i,\pi'(i)}$. Moreover, since both π and π' are products of cycles of the same lengths, $\operatorname{sgn} \pi = \operatorname{sgn} \pi'$. Thus, the contributions of π and π' cancel out in (1.4). (Note that for this part of the argument, we do not need that $\vec{G}$ is Pfaffian.) Thus we may pair up permutations with odd cycles so that they cancel each other.

Now consider a permutation π which consists only of even length cycles and does not vanish in (1.4). As remarked above, π corresponds to an even cycle cover of $\overleftrightarrow{G}$, which, by Lemma 1.12, corresponds to a pair of perfect matchings in G. Because $\vec{G}$ is Pfaffian, each cycle C_j corresponding to a cycle γ_j of π is oddly oriented by $\vec{G}$. Thus, each γ_j contributes a factor -1 to $\prod_{i=0}^{n-1} a_{i,\pi(i)}$ while it also contributes a factor -1 to $\operatorname{sgn} \pi$, being an even cycle. Therefore, overall, π contributes 1 to the sum (1.4). □

Theorem 1.11 provides a polynomial-time algorithm for counting perfect matchings in a graph G, provided G comes equipped with a Pfaffian orientation. But which graphs admit a Pfaffian orientation?

Lemma 1.13. *Let $\vec{G}$ be a connected planar digraph, embedded in the plane. Suppose every face, except the (outer) infinite face, has an odd number of edges that are oriented clockwise. Then, in any simple cycle C, the number of edges oriented clockwise is of opposite parity to the number of vertices of $\vec{G}$ inside C. In particular, $\vec{G}$ is Pfaffian.*

Proof. First, let's see why the condition on simple cycles implies $\vec{G}$ is Pfaffian. Consider a cycle C created by the union of a pair of perfect matchings in G. Then

[1] The sign of π is $+1$ if the cycle decomposition of π has an even number of even length cycles, and -1 otherwise.

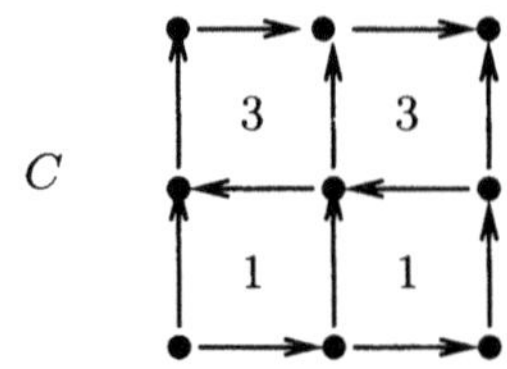

Figure 1.3: Example graph illustrating various quantities in the proof.

C has an even number of vertices inside it, since otherwise there would be a vertex inside C which is matched with a vertex outside C, contradicting planarity. Thus, the number of edges in C oriented clockwise is odd, implying that $\vec{G}$ is Pfaffian.

We now prove the main part of the lemma. Take a cycle C. We need the following definitions:

$$\begin{aligned}
v &= \#\text{ vertices inside } C,\\
k &= \#\text{ edges on } C = \#\text{ vertices on } C,\\
c &= \#\text{ edges on } C \text{ oriented clockwise},\\
f &= \#\text{ faces inside } C,\\
e &= \#\text{ edges inside } C,\\
c_i &= \#\text{ clockwise edges on the boundary of face } i \text{ for } i = 0, \dots, f-1.
\end{aligned}$$

In the example graph illustrated in Figure 1.3, the cycle C is denoted in bold face. Here, $v = 1$, $k = 8$, $c = f = e = 4$, and the various c_i are included in the figure.

According to Euler's formula,

$$\underbrace{(v+k)}_{\#\text{ vertices}} + \underbrace{(f+1)}_{\#\text{ faces}} - \underbrace{(e+k)}_{\#\text{ edges}} = 2,$$

which implies

$$e = v + f - 1. \tag{1.5}$$

Now, for all i, by assumption, $c_i \equiv 1 \pmod 2$, and thus $f \equiv \sum_{i=0}^{f-1} c_i \pmod 2$. On the other hand, $\sum_{i=0}^{f-1} c_i = c + e$, since each interior edge borders two faces, and in exactly one of these it is oriented clockwise. So,

$$\begin{aligned}
f &\equiv c + e\\
&\equiv c + v + f - 1 \pmod 2 \qquad \text{by (1.5)},
\end{aligned}$$

and hence $c + v$ is odd. □

Theorem 1.14. *Every planar graph has a Pfaffian orientation.*

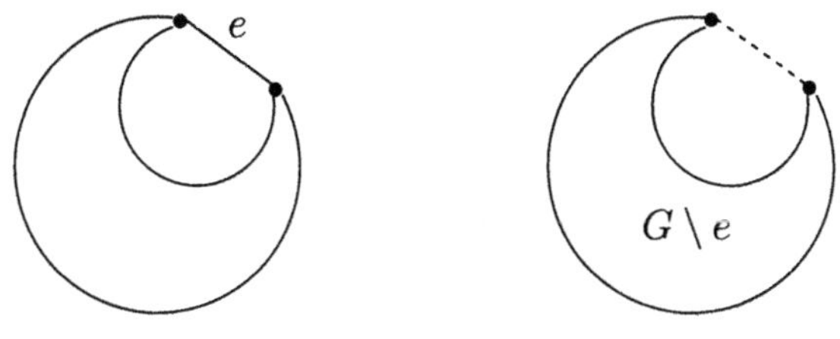

Figure 1.4: Orient e according to the condition of Lemma 1.13.

Proof. Without loss of generality, we may assume G is connected, since we may otherwise treat each connected component separately. We prove the theorem by induction on m, the number of edges. As the base of our induction we take the case when G is a tree, and any orientation is Pfaffian. Now, look at a planar graph G with $m \geq n$ edges, and fix an edge e on the exterior (i.e., e borders the infinite face of G). By the induction hypothesis, $G \setminus e$ has a Pfaffian orientation. Adding e creates just one more face; orient e in such a way that this face has an odd number of edges oriented clockwise. (Figure 1.4 illustrates the situation.) Then, by Lemma 1.13, the orientation is Pfaffian. □

Open Problem. The computational complexity of deciding, for an arbitrary input graph G, whether G has a Pfaffian orientation is open. It is neither known to be in P nor to be NP-complete. The restriction of this decision problem to *bipartite* graphs was recently shown to be decidable by Robertson, Seymour and Thomas [55], and independently by McCuaig.

Note however, that the proof of Theorem 1.14 gives us a polynomial algorithm for finding a Pfaffian orientation of a planar graph G, and hence for counting the number of perfect matchings in G.

Exercise 1.15. In the physics community, perfect matchings are sometimes known as "dimer covers." It is of some interest to know the number of dimer covers of a graph G when G has a regular structure that models, for example, a crystal lattice. Let Λ to be the $L \times L$ square lattice, with vertex set $V(\Lambda) = \{(i,j) : 0 \leq i,j < L\}$ and edge set $E(\Lambda) = \big\{\{(i,j),(i',j')\} : |i-i'| + |j-j'| = 1\big\}$. Exhibit a (nicely structured!) Pfaffian orientation of Λ.

Exercise 1.16. Exhibit a non-planar graph that admits a Pfaffian orientation.

Exercise 1.17. Exhibit a (necessarily non-planar) graph that does not admit a Pfaffian orientation.

Exercise 1.18. The dimer model is one model from statistical physics; another is the Ising model. Computing the "partition function" of an Ising system with underlying graph G in the absence of an external field is essentially equivalent to counting "closed subgraphs" of G: subgraphs $(V, A \subseteq E)$ such that the degree of every vertex $i \in V$ in (V, A) is even (possibly zero). Show that the problem of counting closed subgraphs in a planar graph is efficiently reducible to counting perfect matchings (or dimer covers) in a derived planar graph. The bottom line is that the Ising model for planar systems with no applied field is computationally feasible.

Valiant observes that in the few instances where a counting problem is known to be tractable, it is generally on account of the problem being reducible to the determinant. All the examples presented in this chapter are of this form. This empirical observation remains largely a mystery, though a couple of results in computational complexity give special status to the determinant. For example, around 1991, various authors (Damm, Toda, Valiant, and Vinay) independently discovered that the determinant of an integer matrix is complete for the complexity class GapL under log-space reduction [49, §6].[2] Although this is certainly an interesting result, it does beg the question: why do natural tractable counting problems tend to cluster together in the class GapL? For a further universality property of the determinant, see Valiant [61, §2].

In the other direction, Colbourn, Provan and Vertigan [16] have discovered an interesting, purely combinatorial approach to at least some of the tractable counting problems on planar graphs. In a sense, their result questions the centrality of the determinant.

[2] A function $f : \Sigma^* \to \mathbb{N}$ is in the class #L if there is a log-space non-deterministic Turing machine M such that the number of accepting computations of M on input x is exactly $f(x)$, for all $x \in \Sigma^*$. A function $g : \Sigma^* \to \mathbb{N}$ is in GapL if it can be expressed as $g = f_1 - f_2$ with $f_1, f_2 \in$ #L.

Chapter 2

#P-completeness

Classical complexity theory is mainly concerned with complexity of decision problems, e.g., "Is a given graph G Hamiltonian?"[1] Formally, a *decision problem* is a predicate $\varphi : \Sigma^* \to \{0,1\}$, where Σ is some finite alphabet in which problem instances are encoded.[2] Thus, $x \in \Sigma^*$ might encode a graph G_x (as an adjacency matrix, perhaps) and $\varphi(x)$ is true iff G_x is Hamiltonian.

The most basic distinction in the theory of computational complexity is between predicates that can be decided in time polynomial in the size $|x|$ of the instance, and those that require greater (often exponential) time. This idea is formalised in the complexity class P of polynomial-time predicates. A predicate φ belongs to the complexity class P (and we say that φ is *polynomial time*) if it can be decided by a deterministic Turing machine in time polynomial in the size of the input; more precisely, there is a deterministic Turing machine T and a polynomial p such that, for every input $x \in \Sigma^*$, T terminates after at most $p(|x|)$ steps, accepting if $\varphi(x)$ is true and rejecting otherwise.[3]

Before proceeding, a few vaguely philosophical remarks addressed to readers who have only a passing acquaintance with computational complexity, with the aim of making the chapter more accessible. One motivation for using a robust class of time bounds (namely, all polynomial functions) in the above definition is to render the complexity class P independent of the model of computation. We ought to be able to substitute any "reasonable" sequential model of computation for the Turing machine T in the definition and end up with the same class P. By *sequential* here, we mean that the model should be able to perform just one atomic computational step in each time unit. The "Extended Church-Turing Thesis" is the assertion that the class P is independent of the model of computation used

[1] A *closed* path in G is one that returns to its starting point; a *simple* path is one in which no vertex is repeated; a *Hamilton cycle* in G is a simple closed path that visits every vertex in G. A graph G is *Hamiltonian* if it contains a Hamilton cycle.

[2] Σ^* denotes the set of all finite sequences of symbols in Σ.

[3] Here, $|x|$ denotes the length of the word x.

to define it. It is a thesis rather than a theorem, because we cannot expect to formalise the condition that the model be "reasonable". The upshot of all this is that the reader unfamiliar with the Turing machine model should mentally replace it by some more congenial model, e.g., that of C programs. (For a more expansive treatment of the fundamentals of machine-based computational complexity, refer to standard texts by Papadimitriou [54] or Garey and Johnson [30].)

The important complexity class NP is usually defined in terms of non-deterministic Turing machines. Indeed, NP stands for "N[ondeterministic] P[olynomial time]". In the interests of accessibility, however, we take an alternative but equivalent approach. We say that a predicate $\varphi : \Sigma^* \to \{0,1\}$ belongs to the class NP iff there exists a polynomial-time "witness-checking" predicate $\chi : \Sigma^* \times \Sigma^* \to \{0,1\}$ and a polynomial p such that, for all $x \in \Sigma^*$,

$$\varphi(x) \iff \exists w \in \Sigma^*.\, \chi(x,w) \wedge |w| \le p(|x|)\,. \tag{2.1}$$

(Since the term "polynomial time" has been defined only for monadic predicates, it cannot strictly be applied to χ. Formally, what we mean here is that there is a polynomial-time Turing machine T that takes an input of the form $x\$y$ — where $x, y \in \Sigma^*$ and $\$ \notin \Sigma$ is a special separating symbol — and accepts iff $\chi(x,y)$ is true. The machine T is required to halt in a number of steps polynomial in $|x\$y|$.)

Example 2.1. Suppose x encodes an undirected graph G, y encodes a subgraph H of G, and $\chi(x,y)$ is true iff y is a Hamilton cycle in G. The predicate χ is easily seen to be polynomial time: one only needs to check that H is connected, that H spans G, and that every vertex of H has degree two. Since χ is clearly a witness-checker for Hamiltonicity, we see immediately that the problem of deciding whether a graph is Hamiltonian is in the class NP. Many "natural" decision problems will be seen, on reflection, to belong to the class NP.

As is quite widely known, it is possible to identify within NP a subset of "NP-complete" predicates which are computationally the "hardest" in NP. Since we shall shortly be revisiting the phenomenon of completeness in the context of the counting complexity class #P, just a rough sketch of how this is done will suffice. The idea is to define a notion of reducibility between predicates — polynomial-time many-one (or Karp) reducibility — that allows us to compare their relative computational difficulty. A predicate φ is *NP-hard* if every predicate in NP is reducible to φ; it is *NP-complete* if, in addition, $\varphi \in$ NP.

Logically, there are two possible scenarios: either P = NP, in which case all predicates in NP are efficiently decidable, or P $\subset$ NP, in which case no NP-complete predicate is decidable in polynomial time. Informally, this dichotomy arises because the complete problems are the hardest in NP; formally, it is because the complexity class P is closed under polynomial-time many-one reducibility. Since the former scenario is thought to be unlikely, NP-completeness provides strong circumstantial evidence for intractability. The celebrated theorem of Cook provides a natural example of an NP-complete predicate, namely deciding whether a propositional formula Φ in CNF has a model, i.e., whether Φ is satisfiable. For convenience, this decision problem is referred to as "SAT".

2.1 The class #P

Now we are interested extending the above framework to *counting problems* — e.g., "How many Hamiltonian cycles does a given graph have?" — which can be viewed as functions $f : \Sigma^* \to \mathbb{N}$ mapping (encodings of) problem instances to natural numbers. The class P must be slightly amended to account for the fact we are dealing with functions with codomain $\mathbb{N}$ rather than predicates. A counting problem $f : \Sigma^* \to \mathbb{N}$ is said to belong to the complexity class[4] FP if it is computable by a deterministic Turing machine transducer[5] in time polynomial in the size of the input. As we saw in Chapter 1 (see Theorems 1.1 and 1.11), the following problems are in FP:

Name. #SPANNINGTREES
Instance. A graph G.
Output. The number of spanning trees in G.

Name. #PLANARPM
Instance. A planar graph G.
Output. The number of perfect matchings in G.

The analogue of NP for counting problems was introduced by Valiant [62]. A counting problem $f : \Sigma^* \to \mathbb{N}$ is said to belong to the complexity class #P if there exist a polynomial-time predicate $\chi : \Sigma^* \times \Sigma^* \to \{0, 1\}$ and a polynomial p such that, for all $x \in \Sigma^*$,

$$f(x) = \big|\{w \in \Sigma^* : \chi(x, w) \land |w| \le p(|x|)\}\big| . \tag{2.2}$$

The problem of counting Hamilton cycles in a graph is in #P by identical reasoning to that used in Example 2.1. The complexity class #P is very rich in natural counting problems. Note that elementary considerations entail FP $\subseteq$ #P.

Now, how could we convince ourselves that a problem f is *not* efficiently solvable? Of course, one possibility would be to *prove* that $f \notin$ FP. Unfortunately, such absolute results are beyond the capabilities of the current mathematical theory. Still, as in the case of decision problems, it is possible to provide persuasive evidence for the intractability of a counting problem, based on the assumption that there is *some* problem in #P that is not computable in polynomial time, i.e., that FP $\neq$ #P.[6] With this in mind, we are going to define a class of "most difficult"

[4] Standing for "F[unction] P[olynomial time]" or something similar.

[5] That is, by a TM with a write-only output tape.

[6] This is clearly the counting analogue of the notorious P $\neq$ NP conjecture. Note, however, that FP $\neq$ #P might hold even in the unlikely event that P = NP!

problems in #P, the so-called #P-complete problems, which have the property that if they are in FP, then #P collapses to FP. In other words, if FP $\subset$ #P then no #P-complete counting problem is polynomial-time solvable. For this purpose, we seem to need a notion of reducibility that is more general than the usual many-one reducibility.

Given functions $f, g : \Sigma^* \to \mathbb{N}$, we say that g is *polynomial-time Turing* (or Cook) *reducible* to f, denoted $g \leq_{\mathrm{T}} f$, if there is a Turing machine with an oracle[7] for f that computes g in time polynomial in the input size. The relation $\leq_{\mathrm{T}}$ is transitive; moreover,

$$f \in \mathrm{FP} \wedge g \leq_{\mathrm{T}} f \Rightarrow g \in \mathrm{FP}. \tag{2.3}$$

A function f is #P-*hard* if every function in #P is Turing reducible to f; it is #P-*complete* if, in addition, $f \in$ #P. Just as with the class NP, we have a dichotomy: either FP $=$ #P or no #P-complete counting problem is polynomial-time solvable. Formally, this follows from (2.3), which expresses the fact that FP is closed under polynomial-time Turing reducibility.

What are examples of #P-complete problems? For one thing, the usual generic reduction of a problem in NP to SAT used to prove Cook's theorem is "parsimonious", i.e., it preserves the number of witnesses (satisfying assignments in the case of SAT). It follows that #SAT is #P-complete:

Name. #SAT

Instance. A propositional formula Φ in conjunctive normal form (CNF).

Output. The number of models of (or satisfying assignments to) Φ.

More generally, it appears that NP-complete decision problems tend to give rise to #P-complete counting problems. To be a little more precise: any polynomial-time witness checking function χ gives rise to an NP decision problem Π via (2.1) and a corresponding counting problem $\#\Pi$ via (2.2). Empirically, whenever the decision problem Π is NP-complete, the corresponding counting problem $\#\Pi$ is #P-complete. Simon [56] lists many examples of this phenomenon, and no counterexamples are known. What he observes is that the existing reductions used to establish NP-completeness of decision problems Π are often parsimonious and hence establish also #P-completeness of the corresponding counting problem $\#\Pi$. When the existing reduction is not parsimonious it can be modified so that it becomes so.

[7] An *oracle for* f is an addition to the Turing machine model, featuring a write-only query tape and a read-only response tape. A query $q \in \Sigma^*$ is first written onto the query tape; when the machine goes into a special "query state" the query and response tapes are both cleared and the response $f(x)$ written to the response tape. The oracle is deemed to produce the response in just one time step. In conventional programming language terms, an oracle is a subroutine or procedure, where we discount the time spent executing the body of the procedure.

Open Problem. Is it the case that for every polynomial-time witness-checking predicate χ, the counting problem $\#\Pi$ is #P-complete whenever the decision problem Π is NP-complete? I conjecture the answer is "no", but resolving the question may be difficult. Note that a negative answer could only reasonably be established relative to some complexity theoretic assumption, since it would entail FP $\subset$ #P. Indeed, if FP were to equal #P then every function in #P would be trivially #P-complete.

2.2 A primal #P-complete problem

What makes the theory of #P-completeness interesting is that the converse to the above conjecture is definitely false; that is, there are #P-complete counting problems $\#\Pi$ corresponding to easy decision problems $\Pi \in$ P. A celebrated example [62] is #BIPARTITEPM, that has an alternative formulation as 0,1-PERM:

> *Name.* #BIPARTITEPM
> *Instance.* A bipartite graph G.
> *Output.* The number of perfect matchings in G.

> *Name.* 0,1-PERM
> *Instance.* A square 0,1-matrix $A = (a_{ij} : 0 \le i, j < n)$.
> *Output.* The *permanent*
>
> $$\operatorname{per} A = \sum_{\sigma \in S_n} \prod_{i=0}^{n-1} a_{i,\sigma(i)}$$
>
> of A. Here, S_n denotes the symmetric group, i.e., the sum is over all $n!$ permutations of $[n]$.

To see the correspondence, suppose, for convenience, that G has vertex set $[n]+[n]$, and interpret A as the adjacency matrix of G; thus $a_{ij} = 1$ if (i,j) is an edge of G and $a_{ij} = 0$ otherwise. Then $\operatorname{per} A$ is just the number of perfect matchings in G. In particular, the following theorem implies that planarity (or some slightly weaker assumption) is crucial for the Kasteleyn result (Theorem 1.11).

Theorem 2.2 (Valiant). *0,1-PERM (equivalently, #BIPARTITEPM) is #P-complete.*

It is clear that 0,1-PERM is in #P: the obvious "witnesses" are permutations σ satisfying $\prod_i a_{i,\sigma(i)} = 1$. To prove #P-hardness, we use a sequence of reductions starting at #EXACT3COVER and going via a couple of auxiliary problems #wBIPARTITEMATCH and #wBIPARTITEPM.

Name. #EXACT3COVER

Instance. A set X together with a collection $T \subseteq \binom{X}{3}$ of unordered triples[8] of X.

Output. The number of subcollections $S \subseteq T$ that cover X without overlaps; that is every element of X should be contained in precisely one triple in S.

Name. #wBIPARTITEMATCH

Instance. A bipartite graph G with edge weights $w : E(G) \to \{1, -1, -\frac{5}{3}, \frac{1}{6}\}$. (Why exactly these weights are used will become clearer in the course of the proof.)

Output. The "total weight" of matchings $p_{\text{match}}(G) = \sum_M w(M)$, where M ranges over *all* matchings in G and the weight of a matching is $w(M) = \prod_{e \in M} w(e)$.

Name. #wBIPARTITEPM

Instance. As for #wBIPARTITEMATCH.

Output. As for #wBIPARTITEMATCH, but with "perfect matchings" replacing "matchings".

Remark 2.3. More generally, we might consider a graph G with edge weighting $w : E(G) \to Z \cup \mathbb{C}$, where Z is a set of indeterminates. In this case the expression $p_{\text{match}}(G) = \sum_M w(M)$ appearing in the definition of #wBIPARTITEMATCH is a polynomial in Z. If every edge is assigned a distinct indeterminate, then $p_{\text{match}}(G)$ is the *matching polynomial* of G, i.e., the generating function for matchings in G.

Since #EXACT3COVER is the counting version of an NP-complete problem, we expect it to be #P-complete via parsimonious reduction.

Fact 2.4. #EXACT3COVER *is* #P*-complete.*

Exercise 2.5. (This exercise is mainly directed to readers with some exposure to computational complexity.) Garey and Johnson [30, §7.3] note Fact 2.4 without proof. Since I am not aware of any published proof, we should maybe pause to provide one. Garey and Johnson's reduction [30, §3.1.2] from 3SAT (the restriction of SAT to formulas with three literals per clause) to EXACT3COVER (actually a special case of EXACT3COVER called "3-dimensional matching") is almost parsimonious. The "truth setting component" is fine (each truth assignment corresponds to exactly one pattern of triples). The "garbage collection component" is also fine (it is not strictly parsimonious, but the number of patterns of triples is independent of the truth assignment, which is just as good). The "satisfaction

[8] I'm not sure if $\binom{X}{3}$ is a standard notation for "the set of all unordered triples from X", but it seems natural enough, given the notation 2^X.

testing component" needs some attention, as the number of patterns of triples depends on the truth assignment. However, with a slight modification, this defect may be corrected. Finally, to do a thorough job, we really ought to modify Garey and Johnson's reduction [30, §3.1.1] from SAT to 3SAT to make it parsimonious too.

In the light of Fact 2.4, Theorem 2.2 will follow from the following series of lemmas:

Lemma A. #EXACT3COVER $\le_T$ #wBIPARTITEMATCH.

Lemma B. #wBIPARTITEMATCH $\le_T$ #wBIPARTITEPM.

Lemma C. #wBIPARTITEPM $\le_T$ #BIPARTITEPM ($\equiv$ 0,1-PERM).

Proof of Lemma A. Our construction is based on the weighted bipartite graph H (depicted in Figure 2.1), where the weights of the edges on the left are as indicated, and the edges labelled a_1, a_2 and a_3 will presently all be assigned weight 1. Initially, however, to facilitate discussion, we assign to these edges distinct indeterminates z_1, z_2 and z_3, respectively.

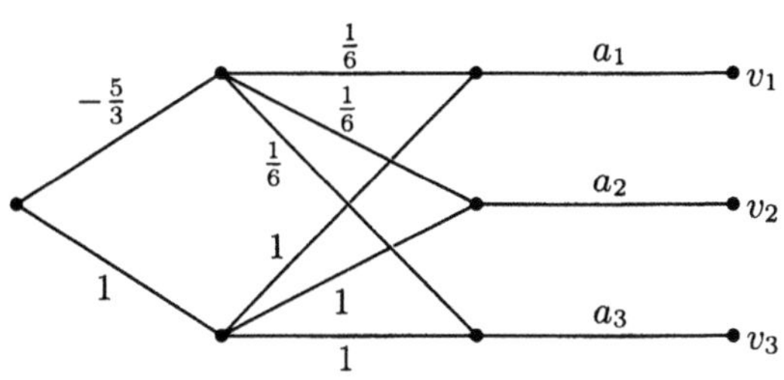

Figure 2.1: The graph H.

By direct computation, the matching polynomial of H, with weights as specified, is

$$p_{\text{match}}(H) = (1 + z_1 z_2 z_3)/3. \tag{2.4}$$

Let us see how to verify (2.4) by calculating the coefficient of $z_1 z_2 z_3$; the other coefficients can be calculated similarly. (Note that there there are only four calculations since, by symmetry, only the *degree* of the monomial is significant.) So suppose we include all three edges a_1, a_2 and a_3, as we must do in order to get a matching that contributes to the coefficient of $z_1 z_2 z_3$. Then we can either add no further edge at all, or add the lower left edge with weight 1, or the upper left edge with weight $-\frac{5}{3}$. Thus, the total weight of such matchings is $(1 + 1 - \frac{5}{3}) z_1 z_2 z_3 = \frac{1}{3} z_1 z_2 z_3$.

Equation (2.4) succinctly expresses the key properties of H that we use. Suppose that H is an (induced) subgraph of a larger graph G, and that H is connected to the rest of G only via the vertices v_1, v_2 and v_3; more precisely, there are no edges of G incident to vertices $V(H) \setminus \{v_1, v_2, v_3\}$ other than the ones depicted. Consider some matching $M' \subseteq E(G) \setminus (E(H) \setminus \{a_1, a_2, a_3\})$ in G, i.e., one that does not use edges from H except perhaps a_1, a_2 and a_3. We call a matching $M \supseteq M'$ in G an *extension* of M' if it agrees with M' on the edge set $E(G) \setminus (E(H) \setminus \{a_1, a_2, a_3\})$. If M' includes all three edges a_i, then the total weight of extensions of M' to a matching M on the whole of G is $\frac{1}{3}w(M')$; a similar claim holds if M' excludes all three edges a_i. In contrast, if M includes some edges a_i and excludes others, then the total weight of extensions of M' is zero. Informally, H acts as a "coordinator" of the three edges a_i.

Using the facts encapsulated in (2.4), we proceed with the reduction of #EXACT3COVER to #wBIPARTITEMATCH. An instance of #EXACT3COVER consists of an underlying set X, and a collection $T \subseteq \binom{X}{3}$ of triples; for convenience set $n := |X|$ and $m := |T|$. We construct a bipartite graph G as follows. Take a separate copy H_t of H for each triple $t = \{\alpha, \beta, \gamma\} \in T$ and label the three pendant edges of H_t with a_α^t, a_β^t, and a_γ^t, respectively. Furthermore, for each $\alpha \in X$, introduce vertices v_α and u_α, and connect them by an edge $\{v_\alpha, u_\alpha\}$ of weight -1. Finally, identify the right endpoint of the edge a_α^t with the vertex v_α whenever $\alpha \in t$ (see Figure 2.2).

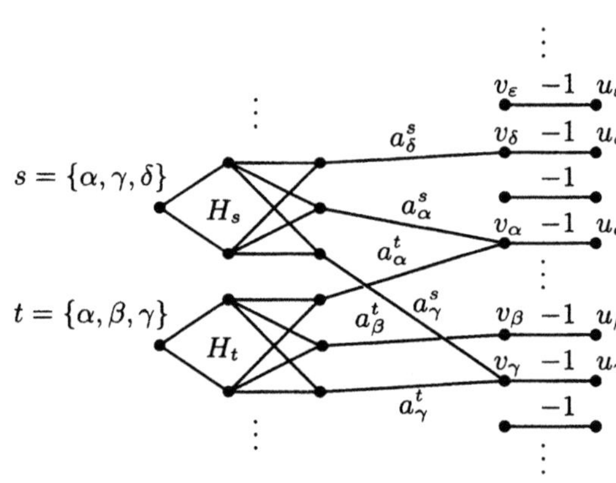

Figure 2.2: A sketch of the graph G.

Recall that the matching polynomial of G is a sum over matchings M in G of the weight $w(M)$ of M. We partition this sum according to the restriction $A = M \cap I$ of M to I, where $I := \{a_\alpha^t : t \in T \wedge \alpha \in t\}$. Computing the total weight of extensions of A to a matching in G is straightforward. For each of the subgraphs H_t, equation (2.4) gives the total weight of extensions of A to that subgraph. For each of the edges $\{v_\alpha, u_\alpha\}$, the total weight of extensions of A to that edge is simply 1 if v_α is covered by A and $(1-1)=0$ otherwise. Expressing these considerations symbolically yields the following expression for the matching polynomial of G:

$$p_{\mathrm{match}}(G) = \sum_M w(M) = \sum_{A\subseteq I} \prod_{t\in T} \varphi_t(A) \prod_{\alpha\in X} \psi_\alpha(A)\,, \tag{2.5}$$

where

$$\varphi_t(A) = \begin{cases} \frac{1}{3}, & \text{if } a^t_\alpha \in A \text{ for all } \alpha \in t; \\ \frac{1}{3}, & \text{if } a^t_\alpha \notin A \text{ for all } \alpha \in t; \\ 0, & \text{otherwise,} \end{cases}$$

and

$$\psi_\alpha(A) = \begin{cases} 1, & \text{if } a^t_\alpha \in A \text{ for some } t \ni \alpha \\ 0, & \text{otherwise.} \end{cases}.$$

Each edge subset A contributing a non-zero term to the sum (2.5) corresponds to an exact 3-cover of X: no element α of X is covered twice (property of a matching), no element of X is uncovered (property of ψ_α), and no triple t is subdivided (property of φ_t). Since every exact 3-cover contributes $(\frac{1}{3})^m$ to (2.5), we obtain

$$p_{\mathrm{match}}(G) = \left(\frac{1}{3}\right)^m \big|\{\text{exact 3-covers of } X \text{ by triples in } T\}\big|. \tag{2.6}$$

Thus, assuming we have an oracle for the left hand side of (2.6), we can compute the number of exact three covers in time polynomial in m and n, and hence polynomial in the size of the #EXACT3COVER instance (X,T). □

Proof of Lemma B. Let G be an instance of #wBIPARTITEMATCH, that is, a bipartite graph with vertex set $V = R \cup B$ and edge set E, where $\binom{R}{2} \cap E = \binom{B}{2} \cap E = \emptyset$, and edge weights from $\{1, -1, \frac{1}{6}, -\frac{5}{3}\}$. Set $r := |R|$ and $b := |B|$, so that $r + b = n := |V|$.

For $0 \le k \le \min\{r, b\}$ (the maximal possible cardinality of a matching in G), we construct a bipartite graph graph G_k as follows. Take a set R' and a set B' of new vertices, $|R'| = b - k$ and $|B'| = r - k$ and connect each vertex in R with each vertex in B' and each vertex in R with each vertex in R' by new edges of weight 1 (see Figure 2.3).

In a similar vein to the proof of Lemma A, we observe that

$$\sum_{\substack{M' \text{ is a perfect} \\ \text{matching in } G_k}} w(M') = (r-k)!\,(b-k)! \sum_{\substack{M \text{ is a } k\text{-} \\ \text{matching in } G}} w(M)\,.$$

Thus, we can compute the total weight of matchings in G by invoking our oracle for #wBIPARTITEPM on every G_k and summing over k. □

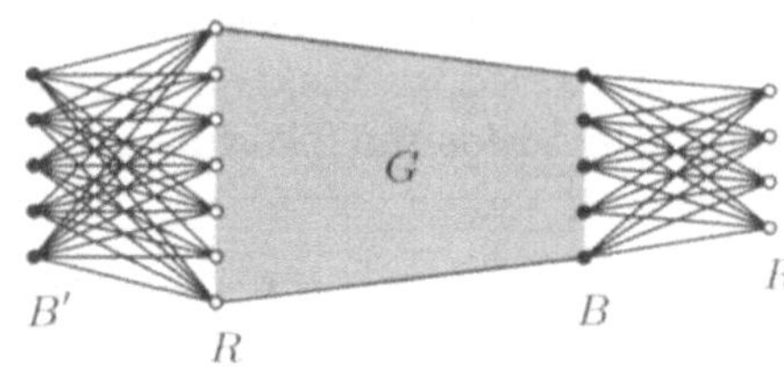

Figure 2.3: The graph G_k.

Proof of Lemma C. Let $G = (V, E)$ be an instance of #BIPARTITEPM, with $|V| = n$. We get rid of the weights one by one using interpolation. Consider a certain weight $\zeta \in \{\frac{1}{6}, -1, -\frac{5}{3}\}$. If we replace it by an indeterminate z, then

$$p(z) := \sum_{\substack{M \text{ is a perfect} \\ \text{matching in } G}} w(M)$$

is a polynomial of degree $d \leq \frac{1}{2}n$. If we can evaluate p at $d+1$ distinct points, say at $k = 1, \ldots, d+1$, we can interpolate to find $p(\zeta)$. (Refer to Valiant [64] for a discussion of efficient interpolation.) In order to find $p(k)$ for fixed k, we construct a graph G_k from G by replacing each edge $\{u, v\}$ of weight z by k disjoint paths of length 3 between u and v such that each edge on these paths has weight 1 (see Figure 2.4).

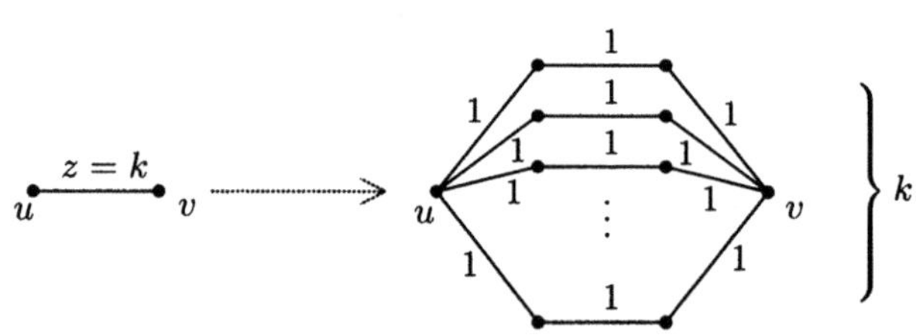

Figure 2.4: Substituting k disjoint paths for an edge.

Then $p(k) = \big|\{\text{perfect matchings in } G_k\}\big|$, and we can determine the right hand side by means of our oracle for #BIPARTITEPM. This completes the proof of the last lemma, and hence of the theorem. □

Remarks 2.6. (a) The intermediate problems in the above proof are not in #P; however, they are "#P-easy", i.e., Turing reducible to a function in #P.

(b) #P-hard counting problems are ubiquitous. In fact, the counting problems in FP are very much the exceptions. The ones we encountered in Chapter 1 — counting trees in directed and undirected graphs (and the related Eulerian circuits in a directed graph), and perfect matchings in a graph (and the related partition function of a planar ferromagnetic Ising system) — are pretty much the only non-trivial examples.

(c) Our reduction from #EXACT3COVER to #BIPARTITEPM used polynomial interpolation in an essential way. Indeed, interpolation features prominently in a majority of #P-completeness proofs. The decision to define #P-completeness with respect to Turing reducibility rather than many-one reducibility is largely motivated by the need to perform many polynomial evaluations (which equate to oracle calls) rather than just one. It is not clear whether the phenomenon of #P-completeness would be as ubiquitous if many-one reducibility were to be used in place of Turing.

(d) Following from the previous observation: Polynomial interpolation is not numerically stable, and does not preserve closeness of approximation. Specifically, we may need to evaluate a polynomial to very great accuracy in order to know some coefficient even approximately. Thus we cannot deduce from the reductions in Lemmas A–C above that *approximating* the permanent is computationally hard, even though approximating #EXACT3COVER is. We exploit this loophole in Chapter 5.

(e) Every problem in NP is trivially #P-easy. It is natural to ask how much bigger #P is than NP. The answer seems to be that it is much bigger. The complexity class PH (Polynomial Hierarchy) is defined similarly to NP, except that arbitrary quantification is allowed in equivalence (2.1), in place of simple existential quantification. PH seems intuitively to be "much bigger" than NP. Yet it is a consequence of Toda's theorem (see [59]) that every problem in PH is #P-easy!

2.3 Computing the permanent is hard on average

While many NP-complete problems are easy to decide on random instances, this does not seem to be the case for counting problems. For example, consider an (imperfect) algorithm $\mathcal{A}$ for computing the permanent of $n \times n$ matrices A over the field $\mathrm{GF}(p)$, for all $n \in \mathbb{N}$ and all primes p, with the following specification:

1. $\mathcal{A}$ has runtime polynomial in n and p;

2. For each n and each p, $\mathcal{A}$ must give the correct result except on some fraction $\frac{1}{3(n+1)}$ of all $n \times n$ matrices over $\mathrm{GF}(p)$.

Theorem 2.7. *No algorithm $\mathcal{A}$ with the above specification exists unless every problem in #P admits a polynomial-time randomised algorithm with low*[9] *error probability.*[10]

Proof. It suffices to show that some particular #P-complete problem, namely 0,1-PERM, admits a polynomial-time randomised algorithm with low error probability. Given an $n \times n$ matrix A with entries from $\{0, 1\}$, if we know $\operatorname{per} A \pmod{p_i}$ for a sequence $p_1, p_2, \ldots, p_n$ of n distinct primes larger than $n+1$, then we can use "Chinese remaindering" to evaluate $\operatorname{per} A$. The method is as follows. If a and b are relatively coprime natural numbers, we can write $1 = ca + db$ with integer coefficients c, d, which can be found by means of the Euclidian algorithm. Now suppose we know the residues $r = x \bmod a$ and $s = x \bmod b$ of an integer x. If we set $y := rdb + sca$, we have $x \equiv y \pmod{a}$ and $x \equiv y \pmod{b}$, and hence $x \equiv y \pmod{ab}$, by relative primality. Thus, inductively, we can compute $(\operatorname{per} A) \bmod p_1 p_2 \ldots p_n$ from the n values $(\operatorname{per} A) \bmod p_i$. But since $\operatorname{per} A$ is a natural number not larger than $n! < p_1 p_2 \ldots p_n$, it is uniquely determined by its residue modulo $p_1 p_2 \ldots p_n$. Moreover, the Prime Number Theorem ensures that we may take the p_i's to be no larger than $O(n \ln n)$; in particular, we can find them by brute force in time polynomial in n.

Thus, it remains to show how, for a fixed prime $n + 2 \le p \le O(n \ln n)$, we can employ $\mathcal{A}$ to compute $(\operatorname{per} A) \bmod p$ with low error probability. For this purpose, we select a matrix R u.a.r. from all $n \times n$ matrices over GF(p). Let z be an indeterminate and consider

$$p(z) := \operatorname{per}(A + zR),$$

regarded as a polynomial of degree at most n with coefficients in GF(p). Using $\mathcal{A}$, we evaluate (in time polynomial in n and p, hence in n) $p(z)$ at $n+1$ points $z = 1, 2, \ldots, n+1$. Observe that, since the numbers $1, \ldots, n+1$ are invertible modulo p, $A+R$, $A+2R, \ldots, A+(n+1)R$ are again random matrices (over GF(p)). Thus, with probability at least $1-(n+1)\frac{1}{3(n+1)} = \frac{2}{3}$, $\mathcal{A}$ will give the correct answer in all instances. Now we interpolate to find $p(0) = (\operatorname{per} A) \bmod p$. □

Remarks 2.8. (a) Feige and Lund [28] have considerably sharpened Theorem 2.7 using techniques from the theory of error-correcting codes.

(b) The property (of a problem) of being as hard on average as in the worst case holds quite generally in high enough complexity classes. Refer to Feigenbaum and Fortnow [29] for a discussion of this phenomenon.

---

[9] We can take "low" to mean $\frac{1}{3}$, since this may be reduced to an arbitrarily small value by repeatedly running $\mathcal{A}$ and taking a majority vote. Note that the error probability decreases to zero exponentially as a function of the number of trials.

[10] The error probability is with respect to random choices made by the algorithm. The input is assumed non-random.

Open Problem. What is the complexity of computing the permanent of a random $0, 1$-matrix? It is reasonable to conjecture that computing the permanent of a $0, 1$-matrix *exactly* is as hard on average as it is in the worst case. However, this purely combinatorial version of the problem leaves no space for the interpolation that was at the heart of the proof of Theorem 2.7.

Chapter 3

Sampling and counting

Accumulated evidence of the kind described in the previous chapter suggests that *exact* counting of combinatorial structures is rarely possible in polynomial time. However, it is in the nature of that evidence[1] that it does not rule out the possibility of *approximate* counting (within arbitrarily small specified relative error). Nor does it rule out the possibility of sampling structures at random from an almost uniform distribution, or even from the precisely uniform distribution (in a suitably defined model of computation), come to that. Indeed these two quests — approximate counting and almost uniform sampling — are intimately related, as we'll see presently.

The aim of this chapter is to illustrate, by means of a concrete example, how almost uniform sampling can be employed for approximate counting, and, after that, how almost uniform sampling can be achieved using Markov chain simulation. But first, let's make precise the various notions we've been talking about informally until now.

3.1 Preliminaries

Consider the problem: given a graph G, return a matching M chosen uniformly at random (u.a.r.) from the set of all matchings in G. In order to discuss sampling problems such as this one we obviously need a model of computation that allows random choices. Less obviously, we also need such a model to discuss approximate counting problems: e.g., given a graph G, compute an estimate of the number of matchings in G that is accurate to within $\pm 10\%$.

A probabilistic Turing machine is a Turing machine T equipped with special coin tossing states. Each coin-tossing state q has two possible successor states q_h and q_t. When T enters state q, it moves on the next step to state q_h with probability $\frac{1}{2}$ and to state q_t with probability $\frac{1}{2}$. Various notions of what it means

[1]Specifically, the property of it described Remark 2.6(d).

for a probabilistic Turing machine to decide a predicate or approximate a function (in each case, with high probability) are possible, leading to various randomised complexity classes.

The probabilistic Turing machine is the usual basis for defining randomised complexity classes, but, more pragmatically, we can alternatively take as our model a random access machine (RAM) equipped with coin-tossing instructions, or a simple programming language that incorporates a random choice statement with two outcomes (themselves statements) that are mutually exclusive and each executed with probability $\frac{1}{2}$. All of these possible models are equivalent, modulo polynomial transformations in run-time. So when the phrase "randomised algorithm" is used in this and subsequent chapters, we are usually free to think in terms of any of the above models. However, when specific time bounds are presented (as opposed to general claims that some algorithm is polynomial time) we shall be taking a RAM or conventional programming language view. For a more expansive treatment of these issues, see Papadimitriou's textbook [54, Chaps 2 & 11].

A *randomised approximation scheme* for a counting problem $f : \Sigma^* \to \mathbb{N}$ (e.g., the number of matchings in a graph) is a randomised algorithm that takes as input an instance $x \in \Sigma^*$ (e.g., an encoding of a graph G) and an error tolerance $\varepsilon > 0$, and outputs a number $N \in \mathbb{N}$ (a random variable of the "coin tosses" made by the algorithm) such that, for every instance x,

$$\Pr\left[e^{-\varepsilon} f(x) \leq N \leq e^{\varepsilon} f(x)\right] \geq \frac{3}{4}. \tag{3.1}$$

We speak of a *fully polynomial randomised approximation scheme*, or *FPRAS*, if the algorithm runs in time bounded by a polynomial in $|x|$ and ε^{-1}.

Remarks 3.1. (a) The number $\frac{3}{4}$ appearing in (3.1) could be replaced by any number in the open interval $(\frac{1}{2}, 1)$.

(b) To first order in ε, the event described in 3.1 is equivalent to $(1-\varepsilon)f(x) \leq N \leq (1+\varepsilon)f(x)$, and this is how the requirement of a "randomised approximation scheme" is more usually specified. However the current definition is equivalent, and has certain technical advantages; specifically, a sequence of approximations of the form $e^{-\varepsilon}\xi_{i+1} \leq \xi_i \leq e^{\varepsilon}\xi_{i+1}$ compose gracefully.

For two probability distributions π and π' on a countable set Ω, define the *total variation distance* between π and π' to be

$$\|\pi - \pi'\|_{\mathrm{TV}} := \frac{1}{2}\sum_{\omega\in\Omega} |\pi(\omega) - \pi'(\omega)| = \max_{A\subseteq\Omega} |\pi(A) - \pi'(A)|. \tag{3.2}$$

A *sampling problem* is specified by a relation $S \subseteq \Sigma^* \times \Sigma^*$ between problem instances x and "solutions" $w \in S(x)$.[2] For example, x might be the encoding of a graph G, and $S(x)$ the set of encodings of all matchings in G. An *almost uniform*

[2] We write $S(x)$ for the set $\{w : x\, S\, w\}$ to avoid awkwardness.

sampler for a solution set $S \subseteq \Sigma^* \times \Sigma^*$ (e.g., the set of all matchings in a graph) is a randomised algorithm that takes as input an instance $x \in \Sigma^*$ (e.g., an encoding of a graph G) and an sampling tolerance $\delta > 0$, and outputs a solution $W \in S(x)$ (a random variable of the "coin tosses" made by the algorithm) such that the variation distance between the distribution of W and the uniform distribution on $S(x)$ is at most δ.[3] An almost uniform sampler is *fully polynomial* if it runs in time bounded by a polynomial in x and $\log \delta^{-1}$. We abbreviate "fully polynomial almost uniform sampler" to FPAUS.

Remarks 3.2. (a) The definitions of FPRAS and FPAUS have obvious parallels. Note however that the dependence of the run-time on the "tolerance" (ε or δ, respectively) is very different: polynomial in ε^{-1} versus $\log \delta^{-1}$ respectively. This difference is deliberate. As we shall see, the relative error in the estimate for $f(x)$ can be improved only at great computational expense, whereas the sampling distribution on $S(x)$ can be made very close to uniform relatively cheaply.

(b) For simplicity, the definitions have be specialised to the case of a uniform distribution on the solution set $S(x)$. However, one could easily generalise the notion of "almost uniform sampler" to general distributions.

The "witness checking predicate" view of the classes NP and #P presented in Chapter 2 carries across smoothly to sampling problems. A witness checking predicate $\chi \subseteq \Sigma^* \times \Sigma^*$ and polynomial p define a sampling problem $S \subseteq \Sigma^* \times \Sigma^*$ via

$$S(x) = \{w \in \Sigma^* : \chi(x, w) \wedge |w| \leq p(|x|)\}, \tag{3.3}$$

where particular attention focuses on polynomial-time predicates χ (c.f. (2.1) and (2.2)). If χ is the "Hamilton cycle" checker of Chapter 2, then the related sampling problem $S(x)$ is that of sampling almost uniformly at random a Hamilton cycle in the graph G encoded by x. So we see that each combinatorial structure gives rise to a trio of related problems: decision, counting and sampling. Furthermore, the second of these at least may be considered in exact (FP) and approximate (FPRAS) forms.

Remark 3.3. The distinction between exactly and almost uniform sampling seems less crucial, and, in any case, technical complications arise when one attempts to define exactly uniform sampling: think of the problem that arises when $|S(x)| = 3$ and we are using the probabilistic Turing machine as our model of computation (or refer to Sinclair [58]).

[3] If $S(x) = \emptyset$ we allow the almost uniform sampler to return a special undefined symbol $\perp$, otherwise it cannot discharge its obligation.

3.2 Reducing approximate counting to almost uniform sampling

Fix a witness-checking predicate χ and consider the associated counting and sampling problems, $f : \Sigma^* \to \mathbb{N}$ and $S \subseteq \Sigma^* \times \Sigma^*$ defined by (2.2) and (3.3), respectively. It is known — under some quite mild condition on χ termed "self-reducibility," which often holds in practice — that the computational complexity of approximating $f(x)$ and sampling almost uniformly from $S(x)$ are closely related. In particular, f admits an FPRAS if and only if S admits an FPAUS. For full details, refer to Jerrum, Valiant and Vazirani [38]. Here we shall explore this relationship in only one direction (FPAUS implies FPRAS) and then only in the context of a specific combinatorial structure, namely matchings in a graph. This reduces the technical complications while retaining the main ideas.

Let $\mathcal{M}(G)$ denote the set of matchings (of all sizes) in a graph G.

Proposition 3.4. *Let G be a graph with n vertices and m edges, where $m \geq 1$ to avoid trivialities. If there is an almost uniform sampler for $\mathcal{M}(G)$ with run-time bounded by $T(n, m, \varepsilon)$, then there is a randomised approximation scheme for $|\mathcal{M}(G)|$ with run-time bounded by $cm^2\varepsilon^{-2}T(n, m, \varepsilon/6m)$, for some constant c. In particular, if there is an FPAUS for $\mathcal{M}(G)$ then there is an FPRAS for $|\mathcal{M}(G)|$.*

Proof. Denote the postulated almost uniform sampler by $\mathcal{S}$. The approximation scheme proceeds as follows. Given G with $E(G) = \{e_1, \ldots, e_m\}$ (in any order), we consider the graphs $G_i := (V(G), \{e_1, \ldots, e_i\})$ for $0 \leq i \leq m$. Thus, G_{i-1} is obtained from G_i by deleting the edge e_i. The quantity $|\mathcal{M}(G)|$ which we would like to estimate can be expressed as a product

$$|\mathcal{M}(G)| = (\varrho_1 \varrho_2 \cdots \varrho_m)^{-1} \tag{3.4}$$

of ratios

$$\varrho_i := \frac{|\mathcal{M}(G_{i-1})|}{|\mathcal{M}(G_i)|}.$$

(Here we use the fact that $|\mathcal{M}(G_0)| = 1$.) Observe that $\mathcal{M}(G_{i-1}) \subseteq \mathcal{M}(G_i)$ and that $\mathcal{M}(G_i) \setminus \mathcal{M}(G_{i-1})$ can be mapped injectively into $\mathcal{M}(G_{i-1})$ by sending M to $M \setminus \{e_i\}$. Hence,

$$\frac{1}{2} \leq \varrho_i \leq 1. \tag{3.5}$$

We may assume $0 < \varepsilon \leq 1$ and $m \geq 1$. In order to estimate the ϱ_i's, we run our sampler $\mathcal{S}$ on G_i with $\delta = \varepsilon/6m$ and obtain a random matching M_i from $\mathcal{M}(G_i)$. Let Z_i be the indicator variable of the event that M_i is, in fact, in $\mathcal{M}(G_{i-1})$, and set $\mu_i := \mathrm{E}\, Z_i = \Pr[Z_i = 1]$. By choice of δ and the definition of the variation distance,

$$\varrho_i - \frac{\varepsilon}{6m} \leq \mu_i \leq \varrho_i + \frac{\varepsilon}{6m}, \tag{3.6}$$

or, from (3.5),

$$\left(1-\frac{\varepsilon}{3m}\right)\varrho_i \le \mu_i \le \left(1+\frac{\varepsilon}{3m}\right)\varrho_i\,; \tag{3.7}$$

so the sample mean of a sufficiently large number s of independent copies[4] $Z_i^{(1)},\ldots,$ $Z_i^{(s)}$ of the random variable Z_i will provide a good estimate for ϱ_i. Specifically, let $s := \lceil 74\varepsilon^{-2}m\rceil \le 75\varepsilon^{-2}m$, and $\overline{Z}_i := s^{-1}\sum_{j=1}^{s} Z_i^{(j)}$.

Note that $\operatorname{Var} Z_i = \mathrm{E}[(Z_i-\mu_i)^2] = \Pr[Z_i = 1](1-\mu_i)^2 + \Pr[Z_i = 0]\mu_i^2 = \mu_i(1-\mu_i)$ and that inequalities (3.5) and (3.7) imply $\mu_i \ge 1/3$. Thus, $\mu_i^{-2}\operatorname{Var} Z_i = \mu_i^{-1} - 1 \le 2$, and hence

$$\frac{\operatorname{Var}\overline{Z}_i}{\mu_i^2} \le \frac{2}{s} \le \frac{\varepsilon^2}{37m}. \tag{3.8}$$

As our estimator for $|\mathcal{M}(G)|$, we use the random variable

$$N := \left(\prod_{i=1}^{m}\overline{Z}_i\right)^{-1}.$$

Note that $\mathrm{E}[\overline{Z}_1\overline{Z}_2\ldots\overline{Z}_m] = \mu_1\mu_2\ldots\mu_m$, and furthermore

$$\begin{aligned}
\frac{\operatorname{Var}[\overline{Z}_1\overline{Z}_2\ldots\overline{Z}_m]}{(\mu_1\mu_2\ldots\mu_m)^2} &= \frac{\mathrm{E}[\overline{Z}_1^2\overline{Z}_2^2\ldots\overline{Z}_m^2]}{\mu_1^2\mu_2^2\ldots\mu_m^2} - 1 \\
&= \prod_{i=1}^{m}\frac{\mathrm{E}[\overline{Z}_i^2]}{\mu_i^2} - 1 && \text{since r.v.'s } \overline{Z}_i \text{ are independent} \\
&= \prod_{i=1}^{m}\left(1+\frac{\operatorname{Var}\overline{Z}_i}{\mu_i^2}\right) - 1 \\
&\le \left(1+\frac{\varepsilon^2}{37m}\right)^m - 1 && \text{by (3.8)} \\
&\le \exp\left(\frac{\varepsilon^2}{37}\right) - 1 \\
&\le \frac{\varepsilon^2}{36},
\end{aligned}$$

since $e^{x/(k+1)} \le 1 + x/k$ for $0 \le x \le 1$ and $k \in \mathbb{N}^+$. Thus, by Chebychev's Inequality,

$$\left(1-\frac{\varepsilon}{3}\right)\mu_1\mu_2\ldots\mu_m \le \overline{Z}_1\overline{Z}_2\ldots\overline{Z}_m \le \left(1+\frac{\varepsilon}{3}\right)\mu_1\mu_2\ldots\mu_m, \tag{3.9}$$

with probability at least $1-(\varepsilon/3)^{-2}(\varepsilon^2/36) = \frac{3}{4}$. Since $e^{-x/k} \le 1 - x/(k+1)$ for $0 \le x \le 1$ and $k \in \mathbb{N}^+$, we have the following weakening of inequality (3.9):

$$e^{-\varepsilon/2}\mu_1\mu_2\ldots\mu_m \le \overline{Z}_1\overline{Z}_2\ldots\overline{Z}_m \le e^{\varepsilon/2}\mu_1\mu_2\ldots\mu_m.$$

[4] Obtained from s independent runs of $\mathcal{S}$ on G_i.

But from (3.7), using again the fact about the exponential function, we have

$$e^{-\varepsilon/2}\varrho_1\varrho_2\cdots\varrho_m \le \mu_1\mu_2\cdots\mu_m \le e^{\varepsilon/2}\varrho_1\varrho_2\cdots\varrho_m\,,$$

which combined with the previous inequality implies

$$e^{-\varepsilon}\varrho_1\varrho_2\ldots\varrho_m \le \overline{Z}_1\overline{Z}_2\ldots\overline{Z}_m \le e^{\varepsilon}\varrho_1\varrho_2\ldots\varrho_m$$

with probability at least $\frac{3}{4}$. Since $\overline{Z}_1\overline{Z}_2\ldots\overline{Z}_m = N^{-1}$ and $\varrho_1\varrho_2\ldots\varrho_m = |\mathcal{M}(G)|^{-1}$, our estimator N for $|\mathcal{M}(G)|$ satisfies requirement (3.1). Thus the algorithm that computes computes N as above is an FPRAS for $|\mathcal{M}(G)|$.

The run-time of the algorithm is dominated by the number of samples required, which is $sm \le 75\varepsilon^{-2}m^2$, multiplied by the time-per-sample, which is $T(n, m, \varepsilon)$; the claimed time-bound is immediate. □

Exercise 3.5. Prove a result analogous to Proposition 3.4 with (proper vertex) q-colourings of a graph replacing matchings. Assume that the number of colours q is strictly greater than the maximum degree Δ of G. There is no need to repeat all the calculation, which is in fact identical. The key thing is to obtain an inequality akin to (3.5), but for colourings in place of matchings.

In light of the connection between approximate counting and almost uniform sampling, methods for sampling from complex combinatorially defined sets gain additional significance. The most powerful technique known to us is Markov chain simulation.

3.3 Markov chains

We deal exclusively in this section with discrete-time Markov chains on a finite state space Ω. Many of the definitions and claims extend to countable state spaces with only minor complication. In Chapter 6 we shall need to employ Markov chains with continuous state spaces, but the corresponding definitions and basic facts will be left until they are required. See Grimmett and Stirzaker's textbook [33] for a more comprehensive treatment.

A sequence $(X_t \in \Omega)_{t=0}^{\infty}$ of random variables (r.v's) is a *Markov chain* (MC), with state space Ω, if

$$\Pr[X_{t+1} = y \mid X_t = x_t, X_{t-1} = x_{t-1}, \ldots, X_0 = x_0] = \Pr[X_{t+1} = y \mid X_t = x_t], \tag{3.10}$$

for all $t \in \mathbb{N}$ and all $x_t, x_{t-1}, \ldots, x_0 \in \Omega$. Equation (3.10) encapsulates the *Markovian property* whereby the history of the MC prior to time t is forgotten. We deal only with *(time-) homogeneous* MCs, i.e., ones for which the right-hand side of (3.10) is independent of t. In this case, we may write

$$P(x, y) := \Pr[X_{t+1} = y \mid X_t = x],$$

where P is the *transition matrix* of the MC. The transition matrix P describes single-step transition probabilities; the t-step transition probabilities P^t are given inductively by

$$P^t(x,y) := \begin{cases} I(x,y), & \text{if } t = 0; \\ \sum_{y' \in \Omega} P^{t-1}(x,y')P(y',y), & \text{if } t > 0, \end{cases}$$

where I denotes the identity matrix $I(x,y) := \delta_{xy}$. Thus $P^t(x,y) = \Pr[X_t = y \mid X_0 = x]$.

A *stationary distribution* of an MC with transition matrix P is a probability distribution $\pi : \Omega \to [0,1]$ satisfying

$$\pi(y) = \sum_{x \in \Omega} \pi(x)P(x,y).$$

Thus if X_0 is distributed as π then so is X_1 (and hence so is X_t for all $t \in \mathbb{N}$). A finite MC always has at least one stationary distribution. An MC is *irreducible* if, for all $x, y \in \Omega$, there exists a $t \in \mathbb{N}$ such that $P^t(x,y) > 0$; it is *aperiodic* if $\gcd\{t : P^t(x,x) > 0\} = 1$ for all $x \in \Omega$.[5] A (finite-state) MC is *ergodic* if it is both irreducible and aperiodic.

Theorem 3.6. *An ergodic MC has a unique stationary distribution π; moreover the MC tends to π in the sense that $P^t(x,y) \to \pi(y)$, as $t \to \infty$, for all $x \in \Omega$.*

Informally, an ergodic MC eventually "forgets" its starting state. Computation of the stationary distribution is facilitated by the following little lemma:

Lemma 3.7. *Suppose P is the transition matrix of an MC. If the function $\pi' : \Omega \to [0,1]$ satisfies*

$$\pi'(x)P(x,y) = \pi'(y)P(y,x), \quad \textit{for all } x, y \in \Omega, \tag{3.11}$$

and

$$\sum_{x \in \Omega} \pi'(x) = 1,$$

then π' is a stationary distribution of the MC. If the MC is ergodic, then clearly $\pi' = \pi$ is the unique stationary distribution.

Proof. We just need to check that π' is invariant. Suppose X_0 is distributed as π'. Then

$$\Pr[X_1 = y] = \sum_{x \in \Omega} \pi'(x)P(x,y) = \sum_{x \in \Omega} \pi'(y)P(y,x) = \pi'(y).$$

□

[5] In the case of an irreducible MC, it is sufficient to verify the condition $\gcd\{t : P^t(x,x) > 0\} = 1$ for just one state $x \in \Omega$.

Remark 3.8. Condition (3.11) is known as *detailed balance.* An MC for which it holds is said to be *time reversible.* Clearly, Lemma 3.7 cannot be applied to non-time-reversible MCs. This is not a problem in practice, since all the MCs we consider are time reversible. In fact, it is difficult in general to determine the stationary distribution of large non-time-reversible MCs, unless there is some special circumstance, for example symmetry, that can be taken into consideration. Furthermore, all the usual methods for constructing MCs with specified stationary distributions produce time-reversible MCs.

Example 3.9. Here is a natural (time homogeneous) MC whose state space is the set $\mathcal{M}(G)$ of all matchings (of all sizes) in a specified graph $G = (V, E)$. The transition matrix of the MC is defined implicitly, by an experimental trial. Suppose the initial state is $X_0 = M \in \mathcal{M}(G)$. The next state X_1 is the result of the following trial:

1. With probability $\frac{1}{2}$ set $X_1 \leftarrow M$ and halt.
2. Otherwise, select $e \in E(G)$ and set $M' \leftarrow M \oplus \{e\}$.[6]
3. If $M' \in \mathcal{M}(G)$ then $X_1 \leftarrow M'$ else $X_1 \leftarrow M$.

Since the MC is time homogeneous, it is enough to describe the first transition; subsequent transitions follow an identical trial. Step 1 may seem a little unnatural, but we shall often include such a looping transition to avoid a certain technical complication. Certainly its presence ensures that the MC is aperiodic. The MC is also irreducible, since it is possible to reach the empty matching from any state by removing edges (and reach any state from the empty matching by adding edges). Thus the MC is ergodic and has a unique stationary distribution.

Exercise 3.10. Demonstrate, using Lemma 3.7, that the stationary distribution of the MC of Example 3.9 is uniform over $\mathcal{M}(G)$.

Exercise 3.10 and Proposition 3.4, taken together, immediately suggest an approach to estimating the number of matchings in a graph. Simulate the MC on $\mathcal{M}(G)$ for T steps, starting at some fixed state X_0, say $X_0 = \emptyset$, and return the final state X_T. If T is sufficiently large, this procedure will satisfy the requirements of an almost uniform sampler for matchings in G. Then the method of Proposition 3.4 may be used to obtain a randomised approximation scheme for the number of matchings $|\mathcal{M}(G)|$. Whether this approach is feasible depends crucially on the rate of convergence of the MC to stationarity. We shall prove in Chapter 5 that a modification[7] of the MC described in Example 3.9 does in fact come "close" to stationarity in a polynomial number of steps (in the size of the graph G), hence yielding an FPRAS for the number of matchings in a graph.

---

[6]The symbol $\oplus$ denotes symmetric difference.

[7]In fact, by comparing the original and modified MCs [19], one can show that the MC as presented in Example 3.9 also converges in polynomially many steps.

Chapter 4

Coupling and colourings

The outline of our programme is now clear: in order to count (approximately) it is enough to be able to sample (almost) uniformly; in order to sample we may simulate an appropriately defined MC. For this approach to be feasible, however, it is important that the MC in question is "rapidly mixing," i.e., that it converges to near-equilibrium in time polynomial (hopefully of small degree) in the size of the problem instance. Since the state space is usually of exponential size as a function of the problem size — think of the number of matchings in a graph as a function of the size of the graph — this is a distinctly non-trivial requirement. We shall presently formalise the rate of convergence to equilibrium in terms of the "mixing time" of the MC. The classical theory of MCs has little to say concerning non-asymptotic bounds on mixing time, and most of the techniques we use have been specially developed for the task in hand. However, there is one classical device, namely coupling, that can be applied in certain situations. As it is the most elementary approach to bounding mixing times, we study it first.

4.1 Colourings of a low-degree graph

Anil Kumar and Ramesh [3] present persuasive evidence that the coupling argument is not applicable to the MC on matchings that was defined at the end of the previous chapter. We therefore use a somewhat simpler example, namely colourings of a low-degree graph. Let $G = (V, E)$ be an undirected graph, and Q a set of q colours. A (proper) q-colouring of G is a an assignment $\sigma : V \to Q$ of colours to the vertices of G such that $\sigma(u) \neq \sigma(v)$ for all edges $\{u, v\} \in E$. In general, even deciding existence of a q-colouring in G is computationally intractable, so we need to impose some condition on G and q.

Denote by Δ the maximum degree[1] of any vertex in G. Brooks' theorem asserts that a q-colouring exists when $q \geq \Delta$, provided $\Delta \geq 3$ and G does not

[1] The *degree* of a vertex is the number of edges incident at that vertex.

1. Select a vertex $v \in V$, u.a.r.
2. Select a colour $c \in Q \setminus X_0(\Gamma(v))$, u.a.r.
3. $X_1(v) \leftarrow c$ and $X_1(u) \leftarrow X_0(u)$ for all $u \neq v$.

Figure 4.1: Trial defining an MC on q-colourings.

contain $K_{\Delta+1}$ as a connected component [7, 9].[2] The proof of Brooks' theorem is effective, and yields a polynomial-time algorithm for constructing a q-colouring. It is also best possible in the (slightly restricted) sense that there are pairs, for example $q = 3$, $\Delta = 4$, which just fail the condition of the theorem, and for which the problem of deciding q-colourability is NP-complete, even when restricted to graphs of maximum degree Δ. So if we are aiming at an efficient sampling procedure for q-colourings we should certainly assume $q \geq \Delta$. Moreover, to approximate the number of q-colourings using the reduction of Exercise 3.5 we need to assume further that $q > \Delta$. Before we complete the work of this section, we shall need to strengthen this condition still further.

So let $G = (V, E)$ be a graph of maximum degree Δ and let Ω denote the set of all q-colourings of G, for some $q > \Delta$. Denote by $\Gamma(v) = \{u : \{u, v\} \in E(G)\}$ the set of vertices in G that are adjacent to v. Consider the (time-homogeneous) MC (X_t) on Ω whose transitions are defined by the experimental trial presented in Figure 4.1. Here we are considering a colouring as a function $V \to Q$, so $X_0(u)$ denotes the colour of vertex u in the initial state, and $X_0(\Gamma(v)) = \{X_0(u) : u \in \Gamma(v)\}$ denotes the set of all colours applied to neighbours of v. Note that the assumption $q > \Delta$ makes it easy to construct a valid initial state X_0.

Exercises 4.1. 1. Prove that the above MC is irreducible (and hence ergodic) under the (stronger) assumption $q \geq \Delta+2$. Further prove, using Lemma 3.7, that its (unique) stationary distribution is uniform over Ω.

2. [Alan Sokal.] Exhibit a sequence of connected graphs of increasing size, with $\Delta = 4$, such that the above MC fails to be irreducible when $q = 5$. (Hint: as a starting point, construct a "frozen" 5-colouring of the infinite square lattice, i.e., the graph with vertex set $\mathbb{Z} \times \mathbb{Z}$ and edge set $\{(i, j), (i', j') : |i - i'| + |j - j'| = 1\}$. The adjective "frozen" applied to a state is intended to indicate that the only transition available from the state is a loop (with probability 1) to the same state.)

3. Design an MC on q-colourings of an arbitrary graph G of maximum degree Δ that is ergodic, provided only that $q \geq \Delta + 1$. The MC should be easily implementable, otherwise there is no challenge! (Hint: use transitions based on edge updates rather than vertex updates.)

[2] K_r denotes the complete graph on r vertices.

We shall show that (X_t) is rapidly mixing, provided $q \geq 2\Delta + 1$, which we assume from now on. (The reader may be assured that this is the very last time we shall strengthen the lower bound on the number of colours!) This result will provide us with a simple and efficient sampling procedure for q-colourings in low-degree graphs.

Suppose (X_t) is any ergodic MC on countable state space Ω, with transition matrix P and initial state $X_0 = x \in \Omega$. For $t \in \mathbb{N}$, the distribution of X_t (the t step distribution) is naturally denoted $P^t(x, \cdot)$. Let π denote the the stationary distribution of the MC, i.e., the limit of $P^t(x, \cdot)$ as $t \to \infty$. Recall the definition of total variation distance from (3.2). We measure the rate of convergence to stationarity of (X_t) by its *mixing time* (from initial state x):

$$\tau_x(\varepsilon) := \min\left\{t : \|P^t(x, \cdot) - \pi\|_{\mathrm{TV}} \leq \varepsilon\right\}. \tag{4.1}$$

Lemma 4.2. *The total variation distance $\|P^t(x, \cdot) - \pi\|_{\mathrm{TV}}$ of the t-step distribution from stationarity is a non-increasing function of t.*

Exercise 4.3. Prove Lemma 4.2. (A proof is given at the end of the chapter.)

In the light of Lemma 4.2, the following definition of mixing time is equivalent to (4.1):

$$\tau_x(\varepsilon) := \min\left\{t : \|P^s(x, \cdot) - \pi\|_{\mathrm{TV}} \leq \varepsilon, \text{ for all } s \geq t\right\}.$$

In other words, once the total variation distance becomes smaller than ε it stays smaller than ε.

Often we would like to make a statement about mixing time that is independent of the initial state, in which case we take a worst-case view and write

$$\tau(\varepsilon) = \max_{x \in \Omega} \tau_x(\varepsilon);$$

we shall refer to $\tau(\varepsilon)$ simply as the *mixing time*.

Remark 4.4. Sometimes the further simplification of setting ε to some constant, say $\varepsilon = \frac{1}{4}$, is made. The justification for this runs as follows. If τ is the first time t at which $\|P^t(x, \cdot) - \pi\|_{\mathrm{TV}} \leq \frac{1}{4}$, then it can be shown [2, Chap. 2, Lemma 20] that $\|P^{k\tau}(x, \cdot) - \pi\|_{\mathrm{TV}} \leq 2^{-k}$ for every $k \in \mathbb{N}$.

Our aim in the next section is to show that the mixing time $\tau(\varepsilon)$ of the MC on colourings is bounded by a polynomial in n and $\log \varepsilon^{-1}$.

Proposition 4.5. *Suppose G is a graph on n vertices of maximum degree Δ. Assuming $q \geq 2\Delta + 1$, the mixing time $\tau(\varepsilon)$ of the MC of Figure 4.1 is bounded above by*

$$\tau(\varepsilon) \leq \frac{q - \Delta}{q - 2\Delta}\, n \ln\left(\frac{n}{\varepsilon}\right).$$

Taking the instance size n into account is a prominent feature of applications of MCs in computer science, especially as compared with classical Markov chain theory. Observe that Proposition 4.5, combined with Proposition 3.4, implies the existence of an FPRAS for q-colourings in graphs of low enough degree.

Corollary 4.6. *Suppose G is a connected graph of maximum degree Δ, and $q \geq 2\Delta + 1$. Then there is an FPRAS for counting q-colourings in G. Denote by n the number of vertices in G and by m the number of edges. Then the running time of this FPRAS as a function of n, m and the error tolerance ε (regarding Δ and q as fixed) is bounded by $cnm^2\varepsilon^{-2}\max\{\ln(m/\varepsilon), 1\}$ for some constant c.*

4.2 Bounding mixing time using coupling

Coupling as a proof technique was discovered by Doeblin in the 1930s. However, its more recent popularity as a tool for bounding mixing time owes much to Aldous. Actually, we shall be using only a restricted form of coupling, namely Markovian coupling.

We start with a ground (time homogeneous) MC (Z_t) with state space Ω and transition matrix P. A *(Markovian) coupling* for (Z_t) is an MC (X_t, Y_t) on $\Omega \times \Omega$, with transition probabilities defined by:

$$\begin{aligned} \Pr[X_1 = x' \mid X_0 = x, Y_0 = y] &= P(x, x'), \\ \Pr[Y_1 = y' \mid X_0 = x, Y_0 = y] &= P(y, y'). \end{aligned} \tag{4.2}$$

Equivalently, with $\widehat{P} : \Omega^2 \to \Omega^2$ denoting the transition matrix of the coupling,

$$\sum_{y' \in \Omega} \widehat{P}((x, y), (x', y')) = P(x, x'),$$

$$\sum_{x' \in \Omega} \widehat{P}((x, y), (x', y')) = P(y, y').$$

Thus, the sequence of r.v.'s (X_t) viewed in isolation forms an MC with transition matrix P, as does the sequence (Y_t).

The easy way to achieve (4.2) would be to assume independence of (X_t) and (Y_t), i.e., that

$$\widehat{P}((x, y), (x', y')) = P(x, x')P(y, y').$$

But this is not necessary, and for our application not desirable. Instead, we are after some correlation that will tend to bring (X_t) and (Y_t) together (whatever their initial states) so that $X_t = Y_t$ for some quite small t. Note that once $X_t = Y_t$, we can arrange quite easily for X_s to be equal to Y_s, for all $s \geq t$, while continuing to satisfy (4.2): just choose a transition from X_s and let Y_s copy it.

The following simple lemma, which is the basis of the coupling method, was perhaps first made explicit by Aldous [1, Lemma 3.6]; see also Diaconis [18, Chap. 4, Lemma 5].

Lemma 4.7 (Coupling Lemma). *Let (X_t, Y_t) be any coupling, satisfying (4.2), based on a ground MC (Z_t) on Ω. Suppose $t : [0,1] \to \mathbb{N}$ is a function satisfying the condition: for all $x, y \in \Omega$, and all $\varepsilon > 0$*

$$\Pr[X_{t(\varepsilon)} \neq Y_{t(\varepsilon)} \mid X_0 = x, Y_0 = y] \leq \varepsilon.$$

Then the mixing time $\tau(\varepsilon)$ of (Z_t) is bounded above by $t(\varepsilon)$.

Proof. Denote by P the transition matrix of (Z_t). Let $A \subseteq \Omega$ be arbitrary. Let $X_0 = x \in \Omega$ be fixed, and Y_0 be chosen according to the stationary distribution π of (Z_t). For any $\varepsilon \in (0,1)$ and corresponding $t = t(\varepsilon)$,

$$\begin{aligned} P^t(x, A) &= \Pr[X_t \in A] \\ &\geq \Pr[X_t = Y_t \wedge Y_t \in A] \\ &= 1 - \Pr[X_t \neq Y_t \vee Y_t \notin A] \\ &\geq 1 - (\Pr[X_t \neq Y_t] + \Pr[Y_t \notin A]) \\ &\geq \Pr(Y_t \in A) - \varepsilon \\ &= \pi(A) - \varepsilon. \end{aligned}$$

Hence, by the second part of definition (3.2), $\|P^t(x, \cdot) - \pi\|_{\mathrm{TV}} \leq \varepsilon$. □

Remark 4.8. Actually we established the stronger conclusion

$$\|P^t(x, \cdot) - P^t(y, \cdot)\|_{\mathrm{TV}} \leq \varepsilon, \quad \text{for all pairs } x, y \in \Omega.$$

This slightly different notion of l_1-convergence corresponds to a slightly different notion of mixing time. This new mixing time has certain advantages, notably submultiplicativity: see Aldous and Fill [2] for more detail.

Let's now see how these ideas may be applied to the q-colouring MC of Figure 4.1. We need to define a coupling on Ω^2 such that the projections onto the first and second coordinates are faithful copies of the original MC in the sense of (4.2). Moreover, we wish the coupling to *coalesce*, i.e., reach a state where $X_t = Y_t$, as soon as possible. Figure 4.2 presents what seems at first sight to be a reasonable proposal. Note that if you hide the random variable Y_1 then the companion random variable X_1 is distributed exactly as if we had used the trial presented in Figure 4.1. (By symmetry, a similar statement could be made about Y_1.) Thus the coupling condition (4.2) is satisfied.

We have argued that the coupling in Figure 4.2 is correct, but how efficient is it? Intuitively, provided we can arrange for $\Pr[c_x = c_y]$ in step 2 to be large, we ought to reach a state with $X_t = Y_t$ (i.e., coalescence) in not too many steps. The Coupling Lemma will then provide a good upper bound on mixing time. In order to understand what is involved in maximising $\Pr[c_x = c_y]$, the following exercise may be useful.

1. Select a vertex $v \in V$ u.a.r.

2. Select a pair of colours (c_x, c_y) from some joint distribution on

$$\big(Q \setminus X_0(\Gamma(v))\big) \times \big(Q \setminus Y_0(\Gamma(v))\big)$$

that has the "correct" marginal distributions; specifically, the distribution of c_x (respectively c_y) should be uniform over $Q \setminus X_0(\Gamma(v))$ (respectively $Q\setminus Y_0(\Gamma(v))$. This joint distribution will be chosen so as to maximise $\Pr[c_x = c_y]$.

3. Set $X_1(v) \leftarrow c_x$ and $Y_1(v) \leftarrow c_y$.

Figure 4.2: A coupling for the MC on colourings

Exercise 4.9. Suppose that $Q = \{0, 1, \ldots, 6\}$, $X_0(\Gamma(v)) = \{3, 6\}$ and $Y_0(\Gamma(v)) = \{4, 5, 6\}$. Thus the sets of legal colours for v in X_1 and Y_1 are $c_x \in \{0, 1, 2, 4, 5\}$ and $c_y \in \{0, 1, 2, 3\}$, respectively. Construct a joint distribution for (c_x, c_y) such that c_x is uniform on $\{0, 1, 2, 4, 5\}$, c_y is uniform on $\{0, 1, 2, 3\}$, and $\Pr[c_x = c_y] = \frac{3}{5}$. Show that your construction is optimal.

The best that can be done in general is as follows.

Lemma 4.10. *Let U be a finite set, A, B be subsets of U, and Z_a, Z_b be random variables, taking values in U. Then there is a joint distribution for Z_a and Z_b such that Z_a (respectively Z_b) is uniform and supported on A (respectively B) and, furthermore,*

$$\Pr[Z_a = Z_b] = \frac{|A \cap B|}{\max\{|A|, |B|\}}$$

Exercise 4.11. Prove Lemma 4.10 and show that the result is best possible. (Assuming your construction in Exercise 4.9 is reasonably systematic, it should be possible to adapt it to the general situation.)

Remark 4.12. The term "coupling" does not have a precise agreed meaning, but its general sense is the following. A pair or perhaps a larger collection of r.v's is given. A coupling is a joint distribution of the several variables that has the correct marginals — i.e., each r.v., when observed independently of the others, has the correct probability distribution — but, taken together, the variables are seen to be correlated. Usually the correlation aims to "bring the r.v's closer together" in some sense. Lemma 4.10 is a special example of an optimal coupling of two r.v's that Lindvall calls the *γ-coupling* [42, §I.5]. The coupling of MCs, as captured in condition (4.2), is another example of the concept.

We are now well prepared for the main result of the chapter.

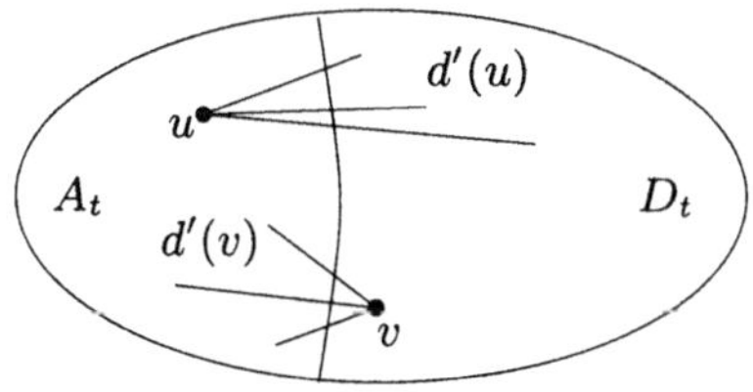

Figure 4.3: Two ways to count the edges spanning the cut (A_t, D_t).

Proof of Proposition 4.5. We analyse the coupling of Figure 4.2 using the joint distribution for the colour-pair (c_x, c_y) that is implicit in Lemma 4.10. Thus, letting

$$\xi := |Q \setminus X_0(\Gamma(v))| \qquad (= \# \text{ legal colours for } v \text{ in } X_1),$$
$$\eta := |Q \setminus Y_0(\Gamma(v))| \qquad (= \# \text{ legal colours for } v \text{ in } Y_1),$$

and

$$\zeta := \big|(Q \setminus X_0(\Gamma(v))) \cap (Q \setminus Y_0(\Gamma(v)))\big| \qquad (= \# \text{ common legal colours}),$$

the probability that the same colour is chosen for both X_1 and Y_1 in step 2 is just

$$\Pr[c_x = c_y] = \frac{\zeta}{\max\{\xi, \eta\}}. \tag{4.3}$$

Consider the situation that obtains after the coupling has been run for t steps. Let $A_t \subseteq V$ be the set of vertices on which the colourings X_t and Y_t agree, and $D_t = V \setminus A_t$ be the set on which they disagree. Let $d'(v)$ denote the number of edges incident at vertex v that have one endpoint in A_t and one in D_t. Clearly,

$$\sum_{v \in A_t} d'(v) = \sum_{u \in D_t} d'(u) = m',$$

where m' is the number of edges of G that span A_t and D_t. (The situation is visualised in Figure 4.3.) We want to prove that the disagreement set D_t tends to get smaller and smaller.

In one transition, the size of the disagreement set D_t changes by at most one. We therefore need to consider just three cases: increasing/decreasing by one or remaining constant. In fact, we just need to compute the probability of the first two, since the third can be got by complementation.

Consider first the probability that $|D_{t+1}| = |D_t| + 1$. For this event to occur, the vertex v selected in step 1 must lie in A_t, and the new colours c_x and c_y selected in step 2 must be different. Observing that the quantities ξ, η and ζ satisfy the linear inequalities

$$\begin{aligned} \xi - \zeta &\le d'(v), \\ \eta - \zeta &\le d'(v), \quad \text{and} \\ \xi, \eta &\ge q - \Delta, \end{aligned} \tag{4.4}$$

we deduce, from (4.3), that

$$\begin{aligned}\Pr[c_x = c_y] &\geq \frac{\max\{\xi,\eta\} - d'(v)}{\max\{\xi,\eta\}} \\ &\geq 1 - \frac{d'(v)}{q - \Delta},\end{aligned}$$

conditional on v being selected in step (1). Thus

$$\begin{aligned}\Pr\big[|D_{t+1}| = |D_t| + 1\big] &= \frac{1}{n}\sum_{v\in A_t} \Pr\big[c_x \neq c_y \mid v \text{ selected}\big] \\ &\leq \frac{1}{n}\sum_{v\in A_t} \frac{d'(v)}{q-\Delta} = \frac{m'}{(q-\Delta)n}. \end{aligned} \tag{4.5}$$

Now consider the probability that $|D_{t+1}| = |D_t| - 1$. For this event to occur, the vertex v selected in step 1 must lie in D_t, and the new colours c_x and c_y selected in step 2 must be the same. The analogues of inequalities (4.4) in this case are

$$\begin{aligned}&\xi - \zeta \leq \Delta - d'(v), \\ &\eta - \zeta \leq \Delta - d'(v), \quad \text{and} \\ &\xi, \eta \geq q - \Delta.\end{aligned}$$

Proceeding as in the previous case,

$$\begin{aligned}\Pr[c_x = c_y] &\geq \frac{\max\{\xi,\eta\} - \Delta + d'(v)}{\max\{\xi,\eta\}} \\ &= 1 - \frac{\Delta - d'(v)}{\max\{\xi,\eta\}} \\ &\geq \frac{q - 2\Delta + d'(v)}{q-\Delta},\end{aligned}$$

conditional on v being selected in step (1). Hence

$$\begin{aligned}\Pr\big[|D_{t+1}| = |D_t| - 1\big] &\geq \frac{1}{n}\sum_{v\in D_t} \frac{q - 2\Delta + d'(v)}{q-\Delta} \\ &\geq \frac{q-2\Delta}{(q-\Delta)n}|D_t| + \frac{m'}{(q-\Delta)n} \end{aligned} \tag{4.6}$$

Define

$$a = \frac{q-2\Delta}{(q-\Delta)n} \quad \text{and} \quad b = b(m') = \frac{m'}{(q-\Delta)n},$$

so that $\Pr\big[|D_{t+1}| = |D_t|+1\big] \leq b$ and $\Pr\big[|D_{t+1}| = |D_t|-1\big] \geq a|D_t|+b$. Provided $a > 0$, i.e., $q > 2\Delta$, the size of the set D_t tends to decrease with t, and hence, intuitively at least, the event $D_t = \emptyset$ should occur with high probability for some $t \le T$ with T not too large. Since $D_t = \emptyset$ is precisely the event that coalescence has occurred, it only remains to confirm this intuition, and quantify the rate at which D_t converges to the empty set. From equations (4.5) and (4.6),

$$\begin{aligned} \mathrm{E}\left[\,|D_{t+1}| \mid D_t\right] &\le b(|D_t|+1) + (a|D_t|+b)(|D_t|-1) \\ &\qquad + (1-a|D_t|-2b)|D_t| \\ &= (1-a)|D_t|. \end{aligned}$$

Thus $\mathrm{E}\,|D_t| \le (1-a)^t |D_0| \le (1-a)^t n$, and, because $|D_t|$ is a non-negative integer random variable, $\Pr[\,|D_t| \ne 0] \le n(1-a)^t \le ne^{-at}$. Note that $\Pr[D_t \ne \emptyset] \le \varepsilon$, provided $t \ge a^{-1}\ln(n\varepsilon^{-1})$, establishing the result. □

Remark 4.13. With a little care, the argument can be pushed to $q = 2\Delta$, though the bound on mixing time worsens by a factor of about n^2. (The r.v. D_t behaves in the boundary case rather like an unbiased random walk, and therefore its expected time to reach the origin $D_t = 0$ is longer; refer, e.g., to Dyer and Greenhill [24], in particular their Theorem 2.1.)

The (direct) coupling technique described here has been used in a number of other applications, such as approximately counting independent sets in a low-degree graph (Luby and Vigoda [46])[3] and estimating the volume of a convex body (Bubley, Dyer and Jerrum [15]).[4] In practice, the versatility of the approach is limited by our ability to design couplings that work well in situations of algorithmic interest. The next section reports on a new technique that promises to extend the effective range of the coupling argument by providing us with a powerful design tool.

4.3 Path coupling

The coupling technique described and illustrated in the previous section is conceptually very simple and appealing. However, in applying the method to concrete situations we face a technical difficulty, which began to surface even in §4.2: how do we encourage (X_t) and (Y_t) to coalesce, while satisfying the demanding constraints (4.2)? Path coupling is an engineering solution to this problem, invented by Bubley and Dyer [11, 12]. Their idea is to define the coupling only on pairs of "adjacent" states, for which the task of satisfying (4.2) is relatively easy, and then to extend the coupling to arbitrary pairs of states by composition of adjacent couplings along a path. The approach is not entirely distinct from classical coupling, and the Coupling Lemma still plays a vital role.

---

[3] Though the subsequent journal article [47] uses the more sophisticated path coupling method, which will be described presently.

[4] The latter application draws inspiration from Lindvall and Rodgers's [43] idea of coupling diffusions by reflection.

1. Select $p \in [n-1]$ according to the distribution f, and $r \in \{0,1\}$ u.a.r.
2. If $r = 1$ and $X_0 \circ (p, p+1) \in \Omega$, then $X_1 := X_0 \circ (p, p+1)$; otherwise, $X_1 := X_0$.

Figure 4.4: Trial defining an MC on linear extensions of a partial order $\prec$.

We illustrate path coupling in the context of a MC on linear extensions of a partial order. We are given a partially ordered set $(V, \prec)$, where $V = [n] = \{0, 1, \ldots, n-1\}$. Denote by $\operatorname{Sym} V$ the symmetric group on V. We are interested in sampling, u.a.r., a member of the set

$$\Omega = \big\{ g \in \operatorname{Sym} V : g(i) \prec g(j) \Rightarrow i \leq j, \text{ for all } i, j \in V \big\}$$

of linear extensions of $\prec$. In forming a mental picture of the the set Ω, the following characterisation may be helpful: $g \in \Omega$ iff the linear order

$$g(0) \sqsubset g(1) \sqsubset \cdots \sqsubset g(n-1) \tag{4.7}$$

extends, or is consistent with, the partial order $\prec$.

As usual, we propose to sample from Ω by constructing an ergodic MC on state space Ω, whose stationary distribution is uniform. The transitions from one linear extension to another are obtained by pre-composing the current linear extension with a random transition $(p, p+1)$. Instead of selecting $p \in [n-1]$ uniformly, we select p from a probability distribution f on $[n-1]$ that gives greater weight to values near the centre of the range. It is possible that this refinement actually reduces the mixing time; in any case, it leads to a simplification of the proof. Formally, the transition probabilities of the MC are defined by the experimental trial presented in Figure 4.4. Note that composition "$\circ$" is to be read right to left, so that (assuming $r = 1$): $X_1(p) = X_0(p+1)$, $X_1(p+1) = X_0(p)$ and $X_1(i) = X_0(i)$, for all $i \notin \{p, p+1\}$.

Provided the probability distribution f is supported on the whole interval $[n-1]$, this MC is irreducible and aperiodic. It is easy to verify, for example using Lemma 3.7, that the stationary distribution of the MC is uniform. As in §3.3, the explicit loop probability of $\frac{1}{2}$ is introduced mainly for convenience in the proof. However, some such mechanism for destroying periodicity is necessary in any case if we wish to treat the empty partial order consistently.

Our analysis of the mixing time of the MC using path coupling will closely follow that of Bubley and Dyer [13]. To apply path coupling, we need first to decide on an adjacency structure for the state space Ω. In this instance we decree that two states g and g' (linear extensions of $\prec$) are adjacent iff $g' = g \circ (i, j)$ for some transposition (i, j) with $0 \leq i < j \leq n-1$; in this case, the distance $d(g, g')$ from g to g' is defined to be $j - i$. Note that the notions of adjacency and distance are symmetric with respect to interchanging g and g', so we can regard this imposed

1. Select $p \in [n-1]$ according to the distribution f, and $r_x \in \{0,1\}$ u.a.r. If $j-i=1$ and $p=i$, set $r_y := 1-r_x$; otherwise, set $r_y := r_x$.
2. If $r_x = 1$ and $X_0 \circ (p, p+1) \in \Omega$ then set $X_1 := X_0 \circ (p, p+1)$; otherwise, set $X_1 := X_0$.
3. If $r_y = 1$ and $Y_0 \circ (p, p+1) \in \Omega$ then set $Y_1 := Y_0 \circ (p, p+1)$; otherwise, set $Y_1 := Y_0$.

Figure 4.5: A possible coupling for the MC on linear extensions.

adjacency structure as a weighted, undirected graph on vertex set Ω; let us refer to this structure as the adjacency graph. It is easily verified that the shortest path in the adjacency graph between two adjacent states is the direct one using a single edge. Thus d may be extended to a metric on Ω by defining $d(g,h)$, for arbitrary states g and h, to be the length of a shortest path from g to h in the adjacency graph.

Next we define the coupling. We need to do this just for adjacent states, as the extension of the coupling via shortest paths to arbitrary pairs of states will be automatic. Suppose that $(X_0, Y_0) \in \Omega^2$ is a pair of states related by $Y_0 = X_0 \circ (i,j)$ for some transposition (i,j) with $0 \le i < j \le n-1$. then the transition to (X_1, Y_1) in the coupling is defined by the experimental trial presented in Figure 4.5. We need to show:

Lemma 4.14. *For adjacent states X_0 and Y_0,*

$$\mathrm{E}\left[d(X_1, Y_1) \mid X_0, Y_0\right] \le \varrho\, d(X_0, Y_0), \tag{4.8}$$

where $\varrho < 1$ is a constant depending on f. For a suitable choice for f, one has $\varrho = 1 - \alpha$, where $\alpha = 6/(n^3 - n)$.

Informally, Lemma 4.14 says that distance between pairs of states in the coupled process tends to decrease: exactly the situation we encountered earlier in the context of the MC on q-colourings. Before proceeding with the proof of Lemma 4.14, let us pause to consider why it is sufficient to establish (4.8) just for adjacent states.

Lemma 4.15. *Suppose a coupling (X_t, Y_t) has been defined, as above, on adjacent pairs of states, and suppose that the coupling satisfies the contraction condition (4.8) on adjacent pairs. Then the coupling can be extended to all pairs of states in such a way that (4.8) holds unconditionally.*

Proof. Suppose $X_0 = x_0 \in \Omega$ and $Y_0 = y_0 \in \Omega$ are now arbitrary. Denote by $P(\cdot,\cdot)$ the transition probabilities of the MC on linear extensions. Let $x_0 = z^{(0)}, z^{(1)}, \dots, z^{(\ell)} = y_0$ be a shortest path from x_0 to y_0 in the adjacency graph.

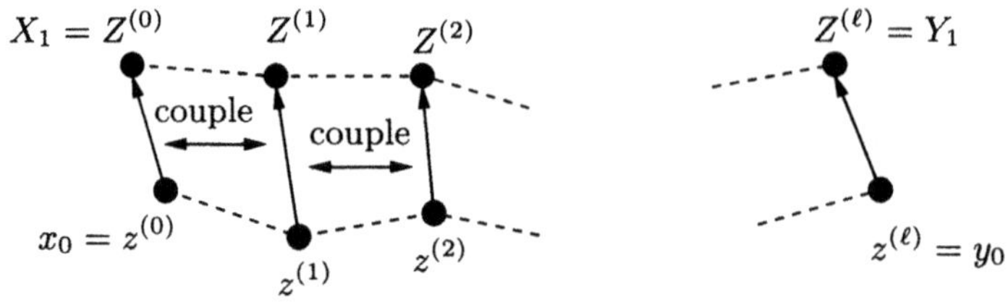

Figure 4.6: Extending a coupling along a shortest path

(Assume a deterministic choice rule for resolving ties.) First select $Z^{(0)} \in \Omega$ according to the probability distribution $P(z^{(0)}, \cdot)$. Now select $Z^{(1)}$ according to the probability distribution induced by the transition $(z^{(0)}, z^{(1)}) \mapsto (Z^{(0)}, Z^{(1)})$ in the coupled process, conditioned on the choice of $Z^{(0)}$; then select $Z^{(2)}$ according to the probability distribution induced by the transition $(z^{(1)}, z^{(2)}) \mapsto (Z^{(1)}, Z^{(2)})$, conditioned on the choice of $Z^{(1)}$; and so on, ending with $Z^{(\ell)}$. (The procedure is visualised in Figure 4.6.)

Let $X_1 = Z^{(0)}$ and $Y_1 = Z^{(\ell)}$. It is routine to verify, by induction on path length ℓ, that Y_1 has been selected according to the (correct) distribution $P(y_0, \cdot)$. Moreover, by linearity of expectation and (4.8),

$$\begin{aligned} \mathrm{E}\left[d(X_1, Y_1) \mid X_0 = x_0, Y_0 = y_0\right] &\le \sum_{i=0}^{\ell-1} \mathrm{E}\, d(Z^{(i)}, Z^{(i+1)}) \\ &\le \varrho \sum_{i=0}^{\ell-1} d(z^{(i)}, z^{(i+1)}) \\ &= \varrho\, d(x_0, y_0). \end{aligned}$$

□

So we see that it is enough to establish the contraction property (4.8) for adjacent pairs of states.

Proof of Lemma 4.14. If $p \notin \{i-1, i, j-1, j\}$ then the tests made in steps (2) and (3) either both succeed or both fail. Thus $Y_1 = X_1 \circ (i, j)$ and $d(X_1, Y_1) = j - i = d(X_0, Y_0)$. Summarising:

$$d(X_1, Y_1) = d(X_0, Y_0), \quad \text{if } p \notin \{i-1, i, j-1, j\}. \tag{4.9}$$

Next suppose $p = i-1$ or $p = j$. These cases are symmetrical, so we consider only the former. With probability at least $\frac{1}{2}$, the tests made in steps (2) and (3) both fail, since $\Pr[r_x = r_y = 0] = \frac{1}{2}$. If this happens, clearly, $d(X_1, Y_1) = j - i = d(X_0, Y_0)$. Otherwise, with probability at most $\frac{1}{2}$, one or other test succeeds. If they both succeed, then

$$\begin{aligned} Y_1 &= Y_0 \circ (i-1, i) \\ &= X_0 \circ (i,j) \circ (i-1,i) \\ &= X_1 \circ (i-1,i) \circ (i,j) \circ (i-1,i) \\ &= X_1 \circ (i-1,j), \end{aligned}$$

and $d(X_1, Y_1) = j - i + 1 = d(X_0, Y_0) + 1$; if only one (say the one in step 2) succeeds, then $Y_1 = Y_0 = X_0 \circ (i,j) = X_1 \circ (i-1,i) \circ (i,j)$, and $d(X_1, Y_1) \le j - i + 1 = d(X_0, Y_0) + 1$. Summarising:

$$\mathrm{E}\left[d(X_1, Y_1) \mid X_0, Y_0, p = i - 1 \vee p = j\right] \le d(X_0, Y_0) + \frac{1}{2}. \tag{4.10}$$

Finally suppose $p = i$ or $p = j-1$. Again, by symmetry, we need only consider the former. There are two subcases, depending on the value of $j - i$. The easier subcase is $j - i = 1$. If $r_x = 1$ then $r_y = 0$ and

$$X_1 = X_0 \circ (i, i+1) = Y_0 \circ (i, i+1) \circ (i, i+1) = Y_0 = Y_1,$$

with a similar conclusion when $r_x = 0$. Thus $d(X_1, Y_1) = 0 = d(X_0, Y_0) - 1$. The slightly harder subcase is the complementary $j - i \ge 2$. The crucial observation is that $X_0 \circ (i, i+1), Y_0 \circ (i, i+1) \in \Omega$ and hence the tests in steps (2) and (3) either both succeed or both fail, depending only on the value of $r_x = r_y$. To see this, observe that

$$X_0(i) \not\prec X_0(i+1) = Y_0(i+1) \not\prec Y_0(j) = X_0(i),$$

from which we may read off the fact that $X_0(i)$ and $X_0(i+1)$ are incomparable in $\prec$. The same argument applies equally to $Y_0(i)$ and $Y_0(i+1)$. If $r_x = 0$ there is no change in state; otherwise, if $r_x = 1$,

$$\begin{aligned} X_1 &= X_0 \circ (i, i+1) \\ &= Y_0 \circ (i,j) \circ (i, i+1) \\ &= Y_1 \circ (i, i+1) \circ (i,j) \circ (i, i+1) \\ &= Y_1 \circ (i+1, j), \end{aligned}$$

and $d(X_1, Y_1) = j - i - 1 = d(X_0, Y_0) - 1$. Summarising both the $j - i = 1$ and $j - i \ge 2$ subcases:

$$\mathrm{E}\left[d(X_1, Y_1) \mid X_0, Y_0, p = i \vee p = j - 1\right] \le e(X_0, Y_0), \tag{4.11}$$

where

$$e(X_0, Y_0) = \begin{cases} 0, & \text{if } d(X_0, Y_0) = 1; \\ d(X_0, Y_0) - \frac{1}{2}, & \text{otherwise.} \end{cases}$$

Note that, in the case $j-i=1$, inequality (4.11) covers just one value of p, namely $p=i=j-1$, instead of two; however, this effect is exactly counterbalanced by an expected reduction in distance of 1 instead of just $\frac{1}{2}$. Combining (4.9)–(4.11) we obtain

$$\mathrm{E}\left[d(X_1,Y_1)\mid X_0,Y_0\right] \le d(X_0,Y_0) - \frac{-f(i-1)+f(i)+f(j-1)-f(j)}{2}.$$

Specialising the probability distribution $f(\cdot)$ to be $f(i):=\alpha(i+1)(n-i-1)$ — where $\alpha := 6/(n^3-n)$ is the appropriate normalising constant — we have, by direct calculation, $-f(i-1)+f(i)+f(j-1)-f(j) = 2\alpha(j-i)$. Since $d(X_0,Y_0)=j-i$, we obtain (4.8) with $\varrho = 1-\alpha$. □

From Lemmas 4.14 and 4.15 it is now a short step to:

Proposition 4.16 (Bubley and Dyer). *The mixing time of the MC on linear extensions (refer to Figure 4.4) is bounded by*

$$\tau(\varepsilon) \le (n^3-n)(2\ln n + \ln \varepsilon^{-1})/6.$$

Proof. By iteration, $\mathrm{E}\left[d(X_t,Y_t)\mid X_0,Y_0\right] \le \varrho^t d(X_0,Y_0)$. For any pair of linear extensions g and h, there is a path in the adjacency graph using only *adjacent* transpositions (i.e., length one edges) that swaps each incomparable pair at most once. Thus $d(X_0,Y_0) \le \binom{n}{2} \le n^2$, and

$$\Pr[X_t \ne Y_t] \le \mathrm{E}\, d(X_t,Y_t) \le (1-\alpha)^t n^2.$$

The latter quantity is less than ε, provided $t \ge (n^3-n)(2\ln n + \ln\varepsilon^{-1})/6$. The result follows directly from Lemma 4.7. □

David Wilson has recently derived a similar $O(n^3 \log n)$ bound on mixing time when f is uniform, i.e, when the transposition $(p, p+1)$ is selected u.a.r.

Exercises 4.17.

1. Use Proposition 4.16 to construct an FPRAS for linear extensions of a partial order.

2. Reprove Proposition 4.5 using path coupling. Note the significant simplification over the direct coupling proof.

New applications of path coupling are regularly being discovered. Bubley, Dyer and Greenhill [14] have presented an FPRAS for q-colourings of a low degree graph that extends the range of applicability of the one described earlier. They were able, for example, to approximate in polynomial time the number of 5-colourings of a graph of maximum degree 3, thus "beating the 2Δ bound" that appeared to exist following the result described in §4.1. Vigoda [66], in a path-coupling tour de force, was able to beat the 2Δ bound uniformly over all sufficiently large Δ; specifically, he proved rapid mixing whenever $q > \frac{11}{6}\Delta$. It is fair to say that neither of these improvements would have been possible without the aid of path coupling.

Dyer and Greenhill have also considered independent sets in a low degree graph [25], and obtained a result similar to, but apparently incomparable with, that of Luby and Vigoda [47]. Bubley and Dyer (again) applied path coupling to establish rapid mixing of a natural Markov chain on sink-free orientations of an arbitrary graph [10]. McShine [50] presents a particularly elegant application of path coupling to sampling tournaments. One further example must suffice: Cooper and Frieze [17] have applied path coupling to analyse the "Swendsen-Wang process," which is commonly used to sample configurations of the "random cluster" or ferromagnetic Potts model in statistical physics.

Finally, for those who skipped Exercise 4.3, here is the missing proof.

Proof of Lemma 4.2. The claim is established by the following sequence of (in-) equalities:

$$
\begin{aligned}
2\,\|P^{t+1}(x,\cdot\,)-\pi\|_{\mathrm{TV}} &= \sum_{y\in\Omega}\bigl|P^{t+1}(x,y)-\pi(y)\bigr| \\
&= \sum_{y\in\Omega}\Bigl|\sum_{z\in\Omega}P^t(x,z)\,P(z,y)-\sum_{z\in\Omega}\pi(z)\,P(z,y)\Bigr| \\
&\le \sum_{y\in\Omega}\sum_{z\in\Omega}\bigl|P^t(x,z)-\pi(z)\bigr|\,P(z,y) \qquad (4.12)\\
&= \sum_{z\in\Omega}\bigl|P^t(x,z)-\pi(z)\bigr|\sum_{y\in\Omega}P(z,y) \\
&= 2\,\|P^t(x,\cdot\,)-\pi\|_{\mathrm{TV}},
\end{aligned}
$$

where (4.12) is the triangle inequality. □

Chapter 5

Canonical paths and matchings

Coupling, at least Markovian coupling, is not a universally applicable method for proving rapid mixing. In this chapter, we define a natural MC on matchings in a graph G and show that its mixing time is bounded by a polynomial in the size of G. Anil Kumar and Ramesh [3] studied a very similar MC to this one, and demonstrated that every Markovian coupling for it takes expected exponential time (in the size of G) to coalesce. In the light of their result, it seems we must take an alternative approach, sometimes called the "canonical paths" method.

5.1 Matchings in a graph

Consider an undirected graph $G = (V, E)$ with vertex set V of size n, and edge set E of size m. Recall that the set of edges $M \subseteq E$ is a *matching* if the edges of M are pairwise vertex disjoint. The vertices that occur as endpoints of edges of M are said to be *covered* by M; the remaining vertices are *uncovered*. For a given graph $G = (V, E)$, we are interested in sampling from the set of matchings of G according to the distribution

$$\pi(M) = \frac{\lambda^{|M|}}{Z} \tag{5.1}$$

where $Z := \sum_M \lambda^{|M|}$, and the sum is over matchings M of all sizes. In statistical physics, the edges in a matching are referred to as "dimers" and the uncovered vertices as "monomers." The probability distribution defined in (5.1) characterises the *monomer-dimer system* specified by G and λ. The normalising factor Z is the *partition function* of the system. The parameter $\lambda \in \mathbb{R}^+$ can be chosen to either favour smaller ($\lambda < 1$) or larger ($\lambda > 1$) matchings, or to generate them from the uniform distribution ($\lambda = 1$).

Note that computing Z exactly is a hard problem. For if it could be done efficiently, one could compute $Z = Z(\lambda)$ at a sequence of distinct values of λ,

1. Select $e = \{u, v\} \in E$ u.a.r.
2. There are three mutually exclusive (but not exhaustive) possibilities:

   ($\uparrow$) If u and v are not covered by X_0, then $M \leftarrow X_0 \cup \{e\}$.

   ($\downarrow$) If $e \in X_0$, then $M \leftarrow X_0 \setminus \{e\}$.

   ($\leftrightarrow$) If u is uncovered and v is covered by some edge $e' \in X_0$ (or vice versa, with the roles of u and v reversed), then $M' \leftarrow M \cup \{e\} \setminus \{e'\}$.

   If none of the above situations obtain, then $M \leftarrow X_0$.
3. With probability $\min\{1, \pi(M)/\pi(X_0)\}$ set $X_1 \leftarrow M$; otherwise, set $X_1 \leftarrow X_0$. (This form of acceptance probability is known as the *Metropolis filter.*)

Figure 5.1: An MC for sampling weighted matchings

and then extract the coefficients of $Z(\lambda)$ by interpolating the computed values. (Observe that $Z(\lambda)$ is a polynomial in λ.) But the highest-order coefficient is just the number of perfect matchings in G. It follows from Theorem 2.2 that evaluating $Z(\lambda)$ at (say) integer points $\lambda \in \mathbb{N}$ is #P-hard. Indeed, with a little more work, one can show that evaluating $Z(\lambda)$ at the particular point $\lambda = 1$ (i.e., counting the number of matchings in G) is #P-complete. Although it is unlikely that Z can be computed efficiently, nothing stops us from having an efficient approximation scheme, in the FPRAS sense of §3.1.

We construct an MC for sampling from distribution (5.1) as shown in Figure 5.1. As usual, denote the state space of the MC by Ω, and its transition matrix by P. Consider two adjacent matchings M and M' with $\pi(M) \leq \pi(M')$. By *adjacent* we just mean that $P(M, M') > 0$, which is equivalent to $P(M', M) > 0$. The transition probabilities between M and M' may be written

$$\begin{aligned} P(M, M') &= \frac{1}{m}, \quad \text{and} \\ P(M', M) &= \frac{1}{m}\frac{\pi(M)}{\pi(M')}, \end{aligned}$$

giving rise to the symmetric form

$$\pi(M)P(M, M') = \pi(M')P(M'M) = \frac{1}{m}\min\{\pi(M), \pi(M')\}. \tag{5.2}$$

The above equality makes clear that the MC is time-reversible, and that its stationary distribution (appealing Lemma 3.7) is π.

Remarks 5.1. (a) The transition probabilities are easy to compute: since a transition changes the number of edges in the current matching by at most one, the acceptance probability in step 3 is either 1 or $\min\{\lambda, \lambda^{-1}\}$, and it is easy to determine which.

(b) Broder [8] was the first to suggest sampling matching by simulating an appropriate MC. His proposal was to construct an MC whose states are perfect matchings (i.e., covering all the vertices of G) and near-perfect matchings (i.e., leaving exactly two vertices uncovered). The MC on all matchings presented in Figure 5.1 was introduced by Jerrum and Sinclair [36].

(c) Time reversibility is a property of MCs that is frequently useful to us; in particular, as we have seen on several occasions, it permits easy verification of the stationary distribution of the MC. However, we shall not make use of the property in the remainder of the chapter, and all the results will hold in the absence of time reversibility.

5.2 Canonical paths

The key to demonstrating rapid mixing using the "canonical paths" technique lies in setting up a suitable multicommodity flow problem. For any pair $x, y \in \Omega$, we imagine that we have to route $\pi(x)\pi(y)$ units of distinguishable fluid from x to y, using the transitions of the MC as "pipes." To obtain a good upper bound on mixing time we must route the flow evenly, without creating particularly congested pipes. To formalise this, we need a measure for congestion.

For any pair $x, y \in \Omega$, define a canonical path $\gamma_{xy} = (x = z_0, z_1, \dots, z_\ell = y)$ from x to y through pairs (z_i, z_{i+1}) of states adjacent in the MC, and let

$$\Gamma := \{\gamma_{xy} \mid x, y \in \Omega\}$$

be the set of all canonical paths. The *congestion* $\varrho = \varrho(\Gamma)$ of the chain is defined by

$$\varrho(\Gamma) := \max_{t=(u,v)} \Bigg\{ \underbrace{\frac{1}{\pi(u)P(u,v)}}_{(\text{capacity of } t)^{-1}} \underbrace{\sum_{x,y:\, \gamma_{xy} \text{ uses } t} \pi(x)\pi(y)\,|\gamma_{xy}|}_{\text{total flow through } t} \Bigg\}. \tag{5.3}$$

where t runs over all transitions, i.e., all pairs of adjacent states of the chain, and $|\gamma_{xy}|$ denotes the length ℓ of the path γ_{xy}.

We want to show that if ϱ is small then so is the mixing time of the MC. Consider some arbitrary "test" function $f : \Omega \to \mathbb{R}$. The variance of f (with respect to π) is

$$\operatorname{Var}_\pi f := \sum_{x \in \Omega} \pi(x)\big(f(x) - \mathrm{E}_\pi f\big)^2 = \sum_{x \in \Omega} \pi(x) f(x)^2 - (\mathrm{E}_\pi f)^2, \tag{5.4}$$

where

$$\mathrm{E}_\pi f := \sum_{x \in \Omega} \pi(x) f(x).$$

It is often convenient to work with an alternative, possibly less familiar expression for variance, namely

$$\operatorname{Var}_\pi f = \frac{1}{2} \sum_{x,y\in\Omega} \pi(x)\pi(y)\bigl(f(x) - f(y)\bigr)^2. \tag{5.5}$$

Equivalence of (5.4) and (5.5) follows from the following sequence of identities:

$$\begin{aligned}
\frac{1}{2} \sum_{x,y\in\Omega} & \pi(x)\pi(y)\bigl(f(x) - f(y)\bigr)^2 \\
&= \sum_{x,y\in\Omega} \bigl[\pi(x)\pi(y)f(x)^2 - \pi(x)\pi(y)f(x)f(y)\bigr] \\
&= \sum_{x\in\Omega} \pi(x)f(x)^2 \sum_{y\in\Omega} \pi(y) - \sum_{x\in\Omega} \pi(x)f(x) \sum_{y\in\Omega} \pi(y)f(y) \\
&= \sum_{x\in\Omega} \pi(x)f(x)^2 - (\mathrm{E}_\pi f)^2 \\
&= \operatorname{Var}_\pi f.
\end{aligned}$$

The variance $\operatorname{Var}_\pi f$ measures the "global variation" of f over Ω. By contrast, the *Dirichlet form*

$$\mathcal{E}_\pi(f,f) := \frac{1}{2} \sum_{x,y\in\Omega} \pi(x)P(x,y)\bigl(f(x) - f(y)\bigr)^2 \tag{5.6}$$

measures the "local variation" of f with respect to the transitions of the MC. The key result relating the congestion ϱ to local and global variation is the following.

Theorem 5.2 (Diaconis and Stroock; Sinclair). *For any function $f : \Omega \to \mathbb{R}$,*

$$\mathcal{E}_\pi(f,f) \ge \frac{1}{\varrho} \operatorname{Var}_\pi f. \tag{5.7}$$

where $\varrho = \varrho(\Gamma)$ is the congestion, defined in (5.3), with respect to any set of canonical paths Γ.

Remarks 5.3. (a) An inequality such as (5.7), which bounds the ratio of the local to the global variation of a function, is often termed a *Poincaré inequality.*

(b) If the congestion ϱ is small, then high global variation of a function entails high local variation. This in turn entails, as we shall see presently, short mixing time.

Proof of Theorem 5.2. We follow Sinclair [57, Thm. 5] whose proof in turn is inspired by Diaconis and Stroock [20].

$$\begin{aligned}
2\operatorname{Var}_\pi f &= \sum_{x,y\in\Omega} \pi(x)\pi(y)\big(f(x)-f(y)\big)^2 \\
&= \sum_{x,y\in\Omega} \pi(x)\pi(y)\Bigg(\sum_{(u,v)\in\gamma_{xy}} 1\cdot\big(f(u)-f(v)\big)\Bigg)^2 && (5.8)\\
&\le \sum_{x,y\in\Omega} \pi(x)\pi(y)\,|\gamma_{xy}| \sum_{(u,v)\in\gamma_{xy}} \big(f(u)-f(v)\big)^2 && (5.9)\\
&= \sum_{u,v\in\Omega} \sum_{\substack{x,y:\\ (u,v)\in\gamma_{xy}}} \pi(x)\pi(y)\,|\gamma_{xy}|\,\big(f(u)-f(v)\big)^2 \\
&= \sum_{u,v\in\Omega} \big(f(u)-f(v)\big)^2 \sum_{\substack{x,y:\\ (u,v)\in\gamma_{xy}}} \pi(x)\pi(y)\,|\gamma_{xy}| \\
&\le \sum_{u,v\in\Omega} \big(f(u)-f(v)\big)^2\,\pi(u)P(u,v)\,\varrho && (5.10)\\
&= 2\varrho\,\mathcal{E}_\pi(f,f).
\end{aligned}$$

Equality (5.8) is a "telescoping sum," inequality (5.9) is Cauchy-Schwarz, and inequality (5.10) is from the definition of ϱ. □

For the following analysis, we modify the chain by making it "lazy." In each step, the lazy MC stays where it is with probability $\frac{1}{2}$, and otherwise makes the transition specified in Figure 5.1. Formally, the transition matrix of the lazy MC is $P_{\mathrm{zz}} := \frac{1}{2}(I+P)$, where I is the identity matrix. It is straightforward to show that the lazy MC is ergodic if the original MC is, in which case the stationary distribution of the two is identical. (In fact, irreducibility of the original MC is enough to guarantee ergodicity of the lazy MC.)

Exercise 5.4. Verify these claims about the lazy MC.

Remarks 5.5. (a) This laziness doubles the mixing time, but ensures that the eigenvalues of the transition matrix are all non-negative, and avoids possible parity conditions that would lead to the MC being periodic or nearly so. In an implementation, to simulate $2t$ steps of the lazy MC, one would generate a sample T from the binomial distribution $\mathrm{Bin}(2t, \frac{1}{2})$, and then simulate T steps of the original, non-lazy MC. Thus, in practice, efficiency would not be compromised by laziness.

(b) The introduction of the lazy chain may seem a little unnatural. At the expense of setting up a little machinery, it can be avoided by using a continuous-time MC rather than a discrete-time MC as we have done. Some other parts of our development would also become smoother in the continuous-time setting. We shall return to this point at the end of the chapter.

Before picking up the argument, some extra notation will be useful. If f is any function $f : \Omega \to \mathbb{R}$ then $P_{zz}f : \Omega \to \mathbb{R}$ denotes the function defined by

$$[P_{zz}f](x) := \sum_{y\in\Omega} P_{zz}(x,y)f(y).$$

The function $P_{zz}f$ is the "one-step averaging" of f. Similarly, $P_{zz}^t f$, defined in an analogous way, is the "t-step averaging" of f: it specifies the averages of f over t-step evolutions of the MC, starting at each of the possible states. If the MC is ergodic (as here), then $P_{zz}^t f$ tends to the constant function $\mathrm{E}_\pi f$ as $t \to \infty$. (Observe that $\mathrm{E}_\pi(P_{zz}f) = \mathrm{E}_\pi f$ and hence $\mathrm{E}_\pi(P_{zz}^t f) = \mathrm{E}_\pi f$; in other words, t-step averaging preserves expectations.) Thus we can investigate the mixing time of the MC by seeing how quickly $\mathrm{Var}_\pi(P_{zz}^t f)$ tends to 0 as $t \to \infty$. This is the idea we shall now make rigorous.

Theorem 5.6. *For any function $f : \Omega \to \mathbb{R}$,*

$$\mathrm{Var}_\pi(P_{zz}f) \le \mathrm{Var}_\pi f - \frac{1}{2}\mathcal{E}_\pi(f,f). \tag{5.11}$$

Proof. We follow closely Mihail's [51] derivation. Consider the one-step averaging of f with respect to the lazy chain:

$$\begin{aligned}[P_{zz}f](x) &= \sum_{y\in\Omega} P_{zz}(x,y)f(y)\\ &= \frac{1}{2}f(x) + \frac{1}{2}\sum_{y\in\Omega} P(x,y)f(y)\\ &= \frac{1}{2}\sum_{y\in\Omega} P(x,y)\big(f(x)+f(y)\big). \end{aligned} \tag{5.12}$$

For convenience, assume[1] $\mathrm{E}_\pi f = 0$, and hence $\mathrm{E}_\pi(P_{zz}f) = 0$. Then the left-hand side of (5.11) is bounded above as follows:

$$\begin{aligned}\mathrm{Var}_\pi(P_{zz}f) &= \sum_{x\in\Omega} \pi(x)\left\{[P_{zz}f](x)\right\}^2\\ &= \frac{1}{4}\sum_{x\in\Omega}\pi(x)\left(\sum_{y\in\Omega}P(x,y)\big(f(x)+f(y)\big)\right)^2 && (5.13)\\ &\le \frac{1}{4}\sum_{x,y\in\Omega}\pi(x)P(x,y)\big(f(x)+f(y)\big)^2, && (5.14)\end{aligned}$$

[1]Otherwise add or subtract a constant, an operation that leaves unchanged the quantities of interest, namely $\mathrm{Var}_\pi f$, $\mathrm{Var}_\pi(P_{zz}f)$ and $\mathcal{E}_\pi(f,f)$.

where step (5.13) uses (5.12), and step (5.14) relies on the fact that the square of the expectation of a r.v. is no greater than the expectation of its square. To get at the right-hand side of (5.11) we use yet another expression for the variance of f:

$$\begin{aligned}\mathrm{Var}_\pi f &= \frac{1}{2}\sum_{x\in\Omega}\pi(x)f(x)^2 + \frac{1}{2}\sum_{y\in\Omega}\pi(y)f(y)^2\\ &= \frac{1}{2}\sum_{x,y\in\Omega}\pi(x)f(x)^2P(x,y) + \frac{1}{2}\sum_{x,y\in\Omega}\pi(x)P(x,y)f(y)^2\\ &= \frac{1}{2}\sum_{x,y\in\Omega}\pi(x)P(x,y)\left(f(x)^2+f(y)^2\right). \qquad (5.15)\end{aligned}$$

Subtracting (5.14) from (5.15) yields

$$\begin{aligned}\mathrm{Var}_\pi f - \mathrm{Var}_\pi(P_{\mathrm{zz}}f) &\geq \frac{1}{4}\sum_{x,y\in\Omega}\pi(x)P(x,y)\left(f(x)-f(y)\right)^2\\ &= \frac{1}{2}\mathcal{E}_\pi(f,f),\end{aligned}$$

as required. □

Combining Theorem 5.2 and Theorem 5.6 gives:

Corollary 5.7. *For any function $f:\Omega\to\mathbb{R}$,*

$$\mathrm{Var}_\pi(P_{\mathrm{zz}}f) \leq \left(1-\frac{1}{2\varrho}\right)\mathrm{Var}_\pi f,$$

where $\varrho = \varrho(\Gamma)$ is the congestion, defined in (5.3), with respect to any set of canonical paths Γ.

Remark 5.8. The algebraic manipulation in the proof of Theorem 5.6 seems mysterious. The discussion of the continuous-time setting at the end of the chapter will hopefully clarify matters a little.

We can now use Corollary 5.7 to bound the mixing time of the chain, by using a special function f. For a subset $A\subseteq\Omega$ of the state space, we consider its indicator function

$$f(x) := \begin{cases}1, & \text{if } x\in A;\\ 0, & \text{otherwise.}\end{cases}$$

Then we have $\mathrm{Var}_\pi f \leq 1$ and therefore

$$\mathrm{Var}_\pi(P_{\mathrm{zz}}^t f) \leq \left(1-\frac{1}{2\varrho}\right)^t \leq \exp\left\{\frac{-t}{2\varrho}\right\},$$

where $P^t_{zz}f$ is the t-step averaging of f. Fix some starting state $x \in \Omega$ and set

$$t = \left\lceil 2\varrho \left(2 \ln \varepsilon^{-1} + \ln \pi(x)^{-1}\right)\right\rceil .$$

This gives

$$\mathrm{Var}_\pi(P^t_{zz}f) \le \exp\left\{-2\ln \varepsilon^{-1} - \ln \pi(x)^{-1}\right\} = \varepsilon^2 \pi(x).$$

On the other hand,

$$\begin{aligned}\mathrm{Var}_\pi(P^t_{zz}f) &\ge \pi(x)\left([P^t_{zz}f](x) - \mathrm{E}_\pi(P^t_{zz}f)\right)^2 \\ &= \pi(x)\left([P^t_{zz}f](x) - \mathrm{E}_\pi\, f\right)^2,\end{aligned}$$

which implies

$$\varepsilon \ge \left|[P^t_{zz}f](x) - \mathrm{E}_\pi\, f\right| = \left|P^t_{zz}(x, A) - \pi(A)\right|$$

for all A. This in turn means that the total variation distance $\|P^t_{zz}(x, \cdot) - \pi\|_{\mathrm{TV}}$ is bounded by ε, and we obtain the following corollary:

Corollary 5.9. *The mixing time of the lazy MC is bounded by*

$$\tau_x(\varepsilon) \le 2\varrho\left(2\ln \varepsilon^{-1} + \ln \pi(x)^{-1}\right),$$

where $\varrho = \varrho(\Gamma)$ is the congestion, defined in (5.3), with respect to any set of canonical paths Γ.

Remark 5.10. The factor 2 in front of the bound on mixing time is an artifact of using the lazy MC.

5.3 Back to matchings

In the previous section, we saw how a general technique (canonical paths) can be used to bound the Poincaré constant of an MC, and how that constant in turn bounds the mixing time. Let's apply this machinery to the matching chain presented in Figure 5.1. Our ultimate goal is to derive a polynomial upper bound on mixing time:

Proposition 5.11. *The mixing time τ of the MC on matchings of a graph G (refer to Figure 5.1) is bounded by*

$$\tau(\varepsilon) \le nm\bar{\lambda}^2\left(4\ln \varepsilon^{-1} + 2n\ln n + n\,|\ln \lambda|\right),$$

where n and m are the number of vertices and edges of G, respectively, and $\bar{\lambda} = \max\{1, \lambda\}$.

Remark 5.12. It is possible, with a little extra work, to improve the upper bound in Proposition 5.11 by a factor of $\bar{\lambda}$: see Exercise 5.17.

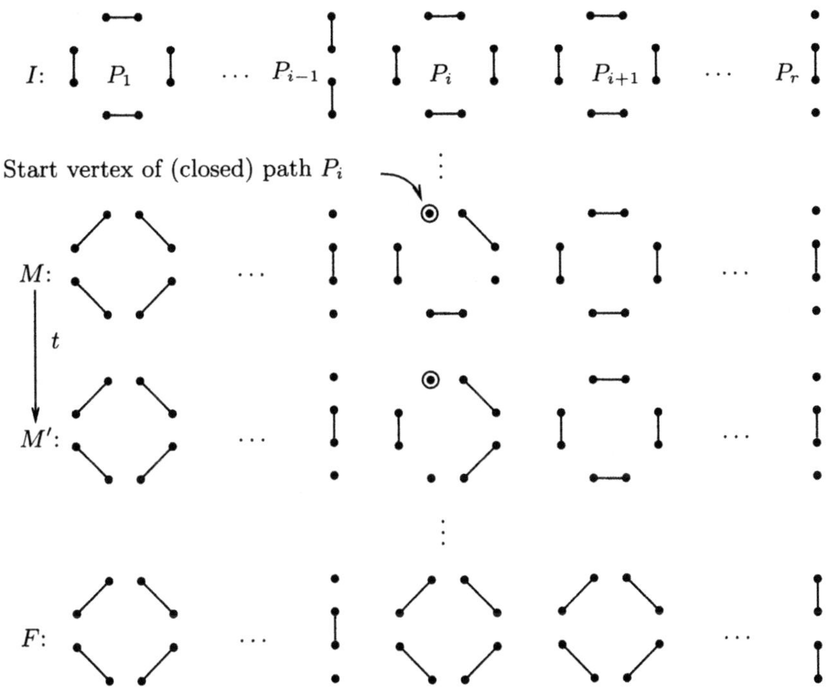

Figure 5.2: A step in a canonical path between matchings

The first step is to define the set Γ of canonical paths. Given two matchings I (initial) and F (final), we need to connect I and F by a canonical path γ_{IF} in the adjacency graph of the matching MC. Along this path, we will have to lose or gain at least the edges in the symmetric difference $I \oplus F$; these edges define a graph of maximum degree two, which decomposes into a collection of paths and even-length cycles, each of them alternating between edges in I and edges in F. If we fix some ordering of the vertices in V, we obtain a unique ordering of the connected components of $(V, I \oplus F)$, by smallest vertex. Within each connected component we may identify a unique "start vertex": in the case of a cycle this will be the smallest vertex, and the case of a path the smaller of the two endpoints. We imagine each path to be oriented away from its start vertex, and each cycle to be oriented so that the edge in I adjacent to the start vertex acquires an orientation away from the start vertex. In Figure 5.2 — which focuses on a particular transition $t = (M, M')$ on the canonical path from I to F — the r connected components of $I \oplus F$ are denoted $P_1, \ldots, P_r$.

To get from I to F, we now process the components of $(V, I \oplus F)$ in the order $P_1, \ldots, P_r$. In each cycle, we first remove the edge in I incident to the start vertex using a $\downarrow$-transition; with a sequence of $\leftrightarrow$-transitions following the cycle's orientation, we then replace I- by F-edges; finally, we perform a $\uparrow$-transitions to add the edge in F incident to the start vertex. In every path, if the start vertex is incident to an F-edge, we use $\leftrightarrow$-transitions along the path and finish by a $\uparrow$-transition in case the path has odd length. If the start vertex is incident to an I-edge, we start with a $\downarrow$-transition, then use $\leftrightarrow$-transitions along the path, and finish with an $\uparrow$-transition in case the path has even length. This concludes the description of the canonical path γ_{IF}. Each transition t on a canonical path γ_{IF} can be thought of as contributing to the processing of a certain connected component of $I \oplus F$; we call this the *current* component (or cycle, or path, if we want to be more specific).

Denote by

$$\mathrm{cp}(t) := \{(I,F) \mid t \in \gamma_{IF}\}$$

the set of pairs $(I,F) \in \Omega$ whose canonical path γ_{IF} uses transition t. To bound the mixing time of the MC, we need to bound from above the congestion

$$\varrho = \max_{t=(M,M')} \left\{ \frac{1}{\pi(M)P(M,M')} \sum_{(I,F)\in \mathrm{cp}(t)} \pi(I)\pi(F)\,|\gamma_{IF}| \right\} \tag{5.16}$$

(c.f. (5.3)), where the maximum is over all transitions $t = (M,M')$. It is not immediately clear how to do this, as the sum is over a set we don't have a ready handle on. Suppose, however, that were able to construct, for each transition $t = (M,M')$, an injective function $\eta_t : \mathrm{cp}(t) \to \Omega$ such that

$$\pi(I)\pi(F) \lessapprox \pi(M)P(M,M')\,\pi(\eta_t(I,F)), \tag{5.17}$$

for all $(I,F) \in \mathrm{cp}(t)$, where the relational symbol "$\lessapprox$" indicates that the left-hand side is larger than the right-hand side by at most a polynomial factor in the "instance size," i.e., some measure of G and λ. Then it would follow that

$$\begin{aligned}
\varrho &\lessapprox \max_t \left\{ \sum_{(I,F)\in \mathrm{cp}(t)} \pi(\eta_t(I,F))\,|\gamma_{IF}| \right\} && \text{from (5.16) and (5.17)} \\
&\lessapprox \max_t \left\{ \sum_{(I,F)\in \mathrm{cp}(t)} \pi(\eta_t(I,F)) \right\} && \text{since } |\gamma_{IF}| \le n \\
&\le 1 && \text{since } \eta_t \text{ is injective.}
\end{aligned}$$

In other words, the congestion ϱ (and hence the mixing time of the MC) is polynomial in the instance size, as we should like.

We now complete the programme by defining an encoding η_t with the appropriate properties, and making exact the calculation just performed. To this end, fix a transition $t = (M,M')$. If t is a $\leftrightarrow$-transition, $(I,F) \in \mathrm{cp}(t)$, and the current component (with respect to the canonical path γ_{IF}) is a cycle, then we say that t

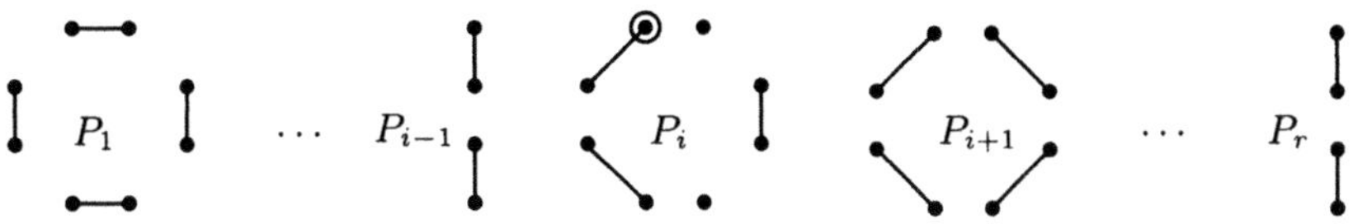

Figure 5.3: The corresponding encoding $\eta_t(X,Y)$.

is *troublesome* (with respect to the path γ_{IF}). If t is troublesome, then we denote by $e_{IFt} \in I$ the (unique) edge in I that is adjacent to the start vertex of the cycle being processed by t. For all $(I,F) \in \mathrm{cp}(t)$, define

$$\eta_t(I,F) = \begin{cases} (I \oplus F \oplus (M \cup M')) \setminus \{e_{IFt}\}, & \text{if } t \text{ is troublesome;} \\ I \oplus F \oplus (M \cup M'), & \text{otherwise.} \end{cases}$$

Roughly speaking, the encoding $C = \eta_t(I,F)$ agrees with I on the components that have been completely processed, and with F on the components that have not been touched yet. Moreover, C agrees with I and F on the edges common to both. (See Figure 5.3.) The crucial properties of η_t are described in the following sequence of claims.

Claim 5.13. *For all transitions t and all pairs $(I,F) \in \mathrm{cp}(t)$, the encoding $C = \eta_t(I,F)$ is a matching; thus η_t is a function with range Ω, as required.*

Proof. Consider the set of edges $A = I \oplus F \oplus (M \cup M')$, and suppose that some vertex, u say, has degree two in A. (Since $A \subseteq I \cup F$, no vertex degree can exceed two.) Then A contains edges $\{u, v_1\}, \{u, v_2\}$ for distinct vertices v_1, v_2, and since $A \subseteq I \cup F$, one of these edges must belong to I and the other to F. Hence both edges belong to $I \oplus F$, which means that neither can belong to $M \cup M'$. Following the form of $M \cup M'$ along the canonical path, however, it is clear that there can be at most one such vertex u; moreover, this happens precisely when t is a troublesome transition and u is the start vertex of the current cycle. Our definition of η_t removes one of the edges adjacent to u in this case, so all vertices in C have degree at most one, i.e., C is indeed a matching. ☐

Claim 5.14. *For every transition t, the function $\eta_t : \mathrm{cp}(t) \to \Omega$ is injective.*

Proof. Let t be a transition, and $(I,F) \in \mathrm{cp}(t)$. We wish to show that the pair (I,F) can be uniquely reconstructed from a knowledge only of t and $\eta_t(I,F)$. It is immediate from the definition of η_t that the symmetric difference $I \oplus F$ can be recovered from $C = \eta_t(I,F)$ using the relation

$$I \oplus F = \begin{cases} (C \oplus (M \cup M')) \cup \{e_{IFt}\}, & \text{if } t \text{ is troublesome;} \\ C \oplus (M \cup M'), & \text{otherwise.} \end{cases}$$

Of course, we don't know, a priori, the identity of the edge e_{IFt}. However, once we have formed the set $C \oplus (M \cup M')$ we can see that e_{IFt} is the unique edge that forms a cycle when added to the current path. There is a slightly delicate issue here: how do we know whether we are in the troublesome case or not? In other words, how to we know whether the current component is a cycle or a path? The answer lies in the convention for choosing the start vertex. It can be checked that choosing the lowest vertex as start vertex leads to a path being oriented in the opposite sense to a cycle in this potentially ambiguous situation.

Given $I \oplus F$, we can at once infer the sequence of paths $P_1, P_2, \ldots, P_r$ that have to be processed along the canonical path from I to F, and the transition t tells us which of these, P_i say, is the current one. The partition of $I \oplus F$ into I and F is now straightforward: I agrees with C on paths $P_1, \ldots, P_{i-1}$, and with M on paths $P_{i+1}, \ldots, P_r$. On the current path, P_i, the matching I agrees with C on the already processed part, and with M on the rest. (If t is troublesome, then the edge e_{IFt} also belongs to I.) Finally, the reconstruction of I and F is completed by noting that $I \cap F = M \setminus (I \oplus F)$, which is immediate from the definition of the paths. Hence I and F can be uniquely recovered from $C = \eta_t(I, F)$, so η_t is injective. □

Claim 5.15. *For all transitions $t = (M, M')$ and all pairs $(I, F) \in \mathrm{cp}(t)$,*

$$\pi(I)\pi(F) \leq m\bar{\lambda}^2 \pi(M) P(M, M')\, \pi(\eta_t(I, F)),$$

where $\bar{\lambda} := \max\{1, \lambda\}$.

Proof. Let $C = \eta_t(I, F)$, and consider the expressions

$$\lambda^{|I|}\lambda^{|F|} \quad \text{and} \quad \lambda^{|M \cup M'|}\lambda^{|C|},$$

which are closely related to the quantities

$$\pi(I)\pi(F) \quad \text{and} \quad \pi(M)P(M, M')\, \pi(\eta_t(I, F))$$

of interest. Each edge $e \in E$ contributes a factor 1, λ or λ^2 to $\lambda^{|I|}\lambda^{|F|}$, according to whether e is in neither, exactly one, or both of I and F. A similar observation can be made about $\lambda^{|M \cup M'|}\lambda^{|C|}$. If $e \notin I$ and $e \notin F$ then $e \notin M \cup M'$ and $e \notin C$, and the contribution to both expressions is 1. If $e \in I$ and $e \in F$ then $e \in M \cup M'$ and $e \in C$ and the contribution to both expressions is λ^2. If $e \in I \oplus F$ then $e \in (M \cup M') \oplus C$ and the contribution to both expressions is λ, with one possible exception: if t is troublesome and $e = e_{IFt}$ then there is a contribution λ to $\lambda^{|I|}\lambda^{|F|}$ and 1 to $\lambda^{|M \cup M'|}\lambda^{|C|}$. Thus,

$$\lambda^{|I|}\lambda^{|F|} \leq \bar{\lambda}\, \lambda^{|M \cup M'|}\lambda^{|C|}.$$

Dividing by Z^2, the square of the partition function, it follows that

$$\pi(I)\pi(F) \leq \bar{\lambda}^2 \pi(M)\pi(C) \quad \text{and} \quad \pi(I)\pi(F) \leq \bar{\lambda}^2 \pi(M')\pi(C),$$

where we have used the fact that $|M|, |M'| \geq |M \cup M'| - 1$. Then

$$\begin{aligned}\pi(I)\pi(F) &\leq \bar{\lambda}^2 \min\{\pi(M), \pi(M')\}\,\pi(C) \\ &= m\bar{\lambda}^2\pi(M)P(M,M')\pi(C) && \text{by (5.2),}\end{aligned}$$

yielding the required inequality. □

Now we are ready to evaluate the congestion ϱ.

Proposition 5.16. *With a set of canonical paths Γ defined as in this section, the congestion $\varrho = \varrho(\Gamma)$ of the MC on matchings of a graph G (refer to Figure 5.1) is bounded by $\varrho \leq nm\bar{\lambda}^2$, where n and m are the number of vertices and edges of G, respectively, and $\bar{\lambda} = \max\{1, \lambda\}$.*

Proof. We just need to make precise the rough calculation following (5.17).

$$\begin{aligned}\varrho &= \max_{t=(M,M')} \left\{ \frac{1}{\pi(M)P(M,M')} \sum_{(I,F)\in\mathrm{cp}(t)} \pi(I)\pi(F)\,|\gamma_{IF}| \right\} \\ &\leq m\bar{\lambda}^2 \sum_{(I,F)\in\mathrm{cp}(t)} \pi(\eta_t(I,F))\,|\gamma_{IF}| && \text{by Claim 5.15} \\ &\leq nm\bar{\lambda}^2 \sum_{(I,F)\in\mathrm{cp}(t)} \pi(\eta_t(I,F)) && \text{since } |\gamma_{IF}| \leq n \\ &\leq nm\bar{\lambda}^2 && \text{by Claim 5.14.}\end{aligned}$$

□

The sought-for bound on mixing time follows immediately.

Proof of Proposition 5.11. Combine Corollary 5.9 and Proposition 5.16, noting the crude bound $\ln \pi(x)^{-1} \leq n \ln n + \frac{1}{2} n\,|\ln \lambda|$, which holds uniformly over $x \in \Omega$. □

Exercise 5.17. Show how to tighten the upper bound in Proposition 5.11 by a factor $\bar{\lambda}$. Since Claim 5.15 is essentially tight when t is troublesome, it is necessary to improve somehow the inequality

$$\sum_{(I,F)\in\mathrm{cp}(t)} \pi(\eta_t(I,F)) \leq 1,$$

by studying carefully the range of η_t. See Jerrum and Sinclair [36], specifically the proof of their Proposition 12.4.

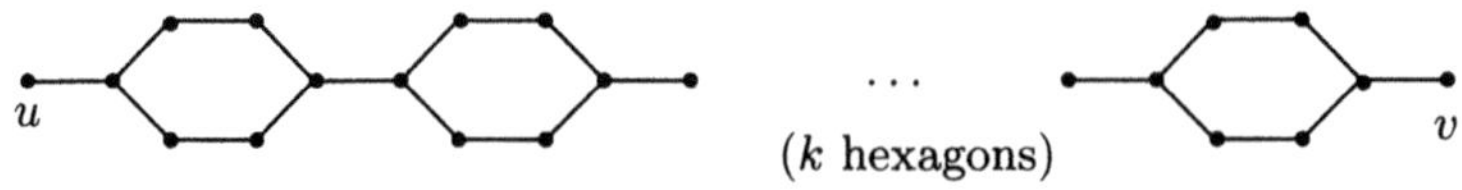

Figure 5.4: A graph with many "near perfect" matchings.

5.4 Extensions and further applications

Let G be a graph with at least one perfect matching (i.e., matching that covers all vertices of G). In the limit, as $\lambda \to \infty$, the partition function $Z(\lambda)$ counts the number of perfect matchings in G. However, the bound on mixing time provided by Proposition 5.11 grows unboundedly with increasing λ, so it is not clear whether the MC we have studied in this chapter provides us with a FPAUS for perfect matchings in G. At first we might hope that it is not necessary to set λ very large; perhaps the distribution (5.1) already places sufficient probability on the totality of perfect matchings at some quite modest λ. (According to Proposition 5.11, we need λ to be bounded by a polynomial in n, the number of vertices in G, to achieve a FPAUS/FPRAS for perfect matchings.)

Unfortunately, there are graphs (see Figure 5.4) for which the perfect matchings make an insignificant contribution to distribution (5.1) unless λ is exponentially large in n. This claim follows from the these easily verified properties of the illustrated graph: (i) it has a unique perfect matching, and (ii) it has 2^k matchings that cover all vertices apart from u and v. The question of whether there exists an FPRAS (equivalently, by the observations of Chapter 3, an FPAUS) for perfect matchings in a general graph is still open at the time of writing. However, progress has been made in some special cases, that of bipartite graphs being perhaps the most interesting.

The problem of counting perfect matchings in a *bipartite* graph is of particular significance, since is is equivalent to evaluating the permanent of a $0,1$-matrix. (Refer to problems #BIPARTITEPM and 0,1-PERM of Chapter 2.) Recently, Jerrum, Sinclair and Vigoda [37] presented an FPRAS for the permanent of a $0,1$-matrix (in fact a general matrix with non-negative entries) using MC simulation. Noting that the counterexample of Figure 5.4 is bipartite, it is clear that we need to introduce a more sophisticated MC to achieve this result. In very rough terms, it is necessary to weight matchings according not just to the *number* of uncovered vertices but also their *locations*. In this way it is possible to access perfect matchings from near-perfect ones via a "staircase" of relatively small steps. Full details may be found in [37].

The canonical paths technique has also been applied by Jerrum and Sinclair to the ferromagnetic Ising model [35] and by Morris and Sinclair to "knapsack solutions" [52]. The latter application is particularly interesting for its use of random canonical paths.

5.5 Continuous time

It is possible to gain a better understanding of Theorem 5.6 and Corollary 5.7 by moving to continuous time.

Associated with any discrete-time MC $(X_t : t \in \mathbb{N})$ is a "continuised" MC $(\widetilde{X}_t : t \in \mathbb{R}^+)$. (We use tilde to distinguish continuous-time notions from their discrete-time analogues.) The MC $(\widetilde{X}_t)$ makes jumps at times $(t_1, t_2, t_3, \ldots)$ where the time increments $t_{i+1} - t_i$, for $i \in \mathbb{N}$, are independent r.v's that are exponentially distributed with mean 1. (Here we use the convention $t_0 = 0$.) Between the jumps, i.e., in the intervals $[t_i, t_{i+1})$, for $i \in \mathbb{N}$, the value of $\widetilde{X}_t$ is constant. The jumps, when they occur, are governed by the same transition matrix P as the original MC (X_t). Informally, we have replaced deterministic time-1 holds between jumps by random, exponential, mean-1 holds. See Norris [53] for a proper treatment of continuous-time MCs.

The continuous-time MC has an "infinitesimal description" $\Pr(\widetilde{X}_{t+dt} = y \mid \widetilde{X}_t = x) = P(x,y)\,dt$ for all $x \neq y$. As a consequence, the distribution of $\widetilde{X}_t$ has a particularly pleasant form:

$$\widetilde{P}^t(x,y) := \Pr(\widetilde{X}_t = y \mid \widetilde{X}_0 = x) = \exp\{(P - I)t\},$$

where I is the identity matrix.[2] As in the discrete-time case, we aim to bound the rate of convergence of $(\widetilde{X}_t)$ to stationarity by analysing the decay of the variance

$$\operatorname{Var}_\pi(\widetilde{P}^t f) := \sum_{x\in\Omega} \pi(x)\big\{[\widetilde{P}^t f](x)\big\}^2, \tag{5.18}$$

where the function $\widetilde{P}^t f : \Omega \to \mathbb{R}$ is defined by

$$[\widetilde{P}^t f](x) := \sum_{y\in\Omega} \widetilde{P}^t(x,y) f(y), \tag{5.19}$$

and $f : \Omega \to \mathbb{R}$ is any test function with $\mathrm{E}_\pi f = 0$.

By calculus, starting with (5.18) and (5.19), we may derive (calculation left to the reader):

$$\frac{d}{dt}\operatorname{Var}_\pi(\widetilde{P}^t f) = 2 \sum_{x,y\in\Omega} \pi(x)\big(P(x,y) - I(x,y)\big)\,[\widetilde{P}^t f](x)\,[\widetilde{P}^t f](y).$$

Hence, setting $t = 0$, we obtain

$$\begin{aligned}
\frac{d}{dt}\operatorname{Var}_\pi(\widetilde{P}^t f)\bigg|_{t=0} &= 2 \sum_{x,y\in\Omega} \pi(x)\big(P(x,y) - I(x,y)\big) f(x) f(y) \\
&= 2 \sum_{x,y\in\Omega} \pi(x) P(x,y) f(x) f(y) - 2\operatorname{Var}_\pi f \\
&= -2\,\mathcal{E}_\pi(f,f),
\end{aligned}$$

[2] The exponential function applied to matrices can be understood as a convergent sum $\exp Q := I + Q + Q^2/2! + Q^3/3! + \cdots$.

a continuous-time analogue of Theorem 5.6.

Applying Theorem 5.2, we see that $\mathrm{Var}_\pi(\widetilde{P}^t f)$ is bounded by the solution of the differential equation $\dot{v} = -(2/\varrho)v$, and hence

$$\mathrm{Var}_\pi(\widetilde{P}^t f) \le \exp\left\{ -\frac{2t}{\varrho} \right\} \mathrm{Var}_\pi f, \tag{5.20}$$

a continuous-time analogue of Corollary 5.7.

Exercise 5.18. Follow through in detail the calculations sketched above.

Remarks 5.19. (a) The rate of decay of variance promised by (5.20) is faster than Corollary 5.7 by a factor 4. A factor 2 is explained by the avoidance of the lazy MC, but the remaining factor 2 is "real." This suggests that the calculation in Theorem 5.6 is not only a little mysterious, but also gives away a constant factor.

(b) Simulating the continuised MC is unproblematic, and can be handled by a device similar to that employed in the case of the lazy MC (c.f. Remarks 5.5). To obtain a sample from the distribution of $\widetilde{X}_t$: (i) generate a sample T from the Poisson distribution with mean t, and then (ii) simulate the discrete-time MC for T steps.

Chapter 6

Volume of a convex body

We arrive at one of the most important applications of the Markov chain Monte Carlo method: the estimation of the volume of a convex body. For a convex body K in low dimensional Euclidean space, say two or three dimensions, it is not too difficult to estimate the volume of K within reasonable relative error using a direct Monte Carlo approach. Depending on how K is presented, it may even be possible to find the volume exactly without too much difficulty. In this chapter, therefore, we imagine the dimension n of the space to be large, and certainly greater than 3.

There are two related problems:

- sample uniformly at random a point from the convex body K;
- estimate the volume $\mathrm{vol}_n K$ of K.

We will first look at the problem of random sampling in K. Since volume is the limit of a sum, it is not surprising, in the light of examples contained in previous chapters, that the second problem can be reduced to the first. We shall look first at the problem of random sampling in K; the reduction of volume estimation to sampling will be covered at the end of the chapter.

The convex body is given as an oracle which, for a point $x \in \mathbb{R}^n$, tells whether or not $x \in K$ (see Figure 6.1). This oracle model subsumes several possible conventions for describing inputs. For example, in the case of a convex polytope defined by a set of linear inequalities it is of course easy to implement the oracle. A convex polytope presented as the convex hull of its vertices it is a little harder, but it can still be done, by linear programming. In some applications, the assumption of an *exact* oracle that accurately decides whether $x \in K$ may be unrealistic. In an implementation we would almost certainly be using arithmetic with bounded precision, and we could not always know for sure whether were in or out. In fact, it is possible to relax the definition of oracle to incorporate some fuzziness at the boundary of K without loosing much algorithmically. One of the many simplifications we shall make in this chapter is to assume exact arithmetic and an exact oracle. For a much fuller picture, refer to Kannan, Lovász and Simonovits [39].

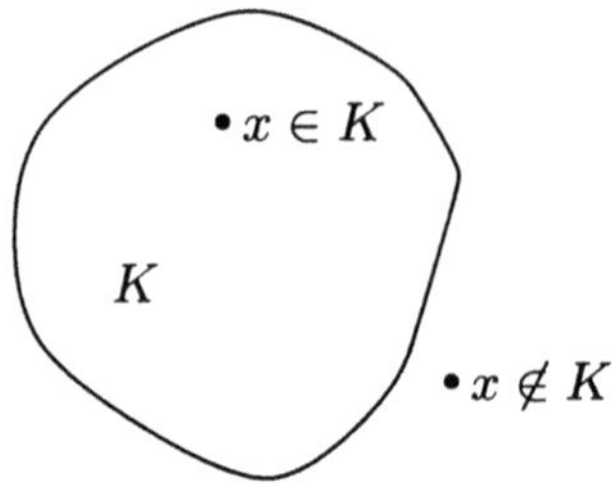

Figure 6.1: Oracle for K.

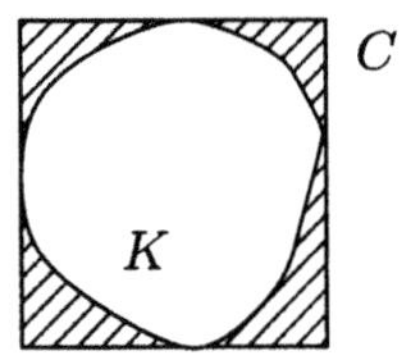

Figure 6.2: Sampling by "direct" Monte Carlo.

The first thing to be noticed in this endeavour is that some intuitively appealing approaches do not work very well. Let us consider a conventional application of the Monte Carlo method to the problem. Say we shrink a box C around K as tightly as possible (see Figure 6.2), sample a point x uniformly at random from C, and return x if $x \in K$; otherwise repeat the sampling if $x \notin K$. This simple idea works well in low dimension, but not in high dimension, where the volume ratio $\operatorname{vol}_n K / \operatorname{vol}_n C$ can be exponentially small. This phenomenon may be illustrated by a very simple example. Let $K = B_n(0,1)$ be the unit ball, and $C = [-1,1]^n$ the smallest enclosing cube. In this instance the ratio in question may be calculated exactly, and is $\operatorname{vol}_n K / \operatorname{vol}_n C = 2\pi^{n/2}/(2^n n\,\Gamma(n/2))$, which decays rapidly with n.[1] In the light of this observation, it seems that a random walk through K may provide a better alternative.

Dyer, Frieze and Kannan [23] were the first to propose a suitable random walk for sampling random points in a convex body K and prove that its mixing time scales as a polynomial in the dimension n. As a consequence, they obtained the first FPRAS for the volume of a convex body. Needless to say, this result was a major breakthrough in the field of randomised algorithms. Their approach was to divide K into a n-dimensional grid of small cubes, with transitions available between cubes sharing a facet (i.e., an $(n-1)$-dimensional face). This proposal imposes a preferred coordinate system on K leading to some technical complications. Here, instead, we use the coordinate-free "ball walk" of Lovász and Simonovits [44].

[1] The Gamma function extends the factorial function to non-integer values. When n is even, $\Gamma(n/2) = (n/2-1)!$, so it is easy to see that the ratio $\operatorname{vol}_n K / \operatorname{vol}_n C$ tends to 0 exponentially fast.

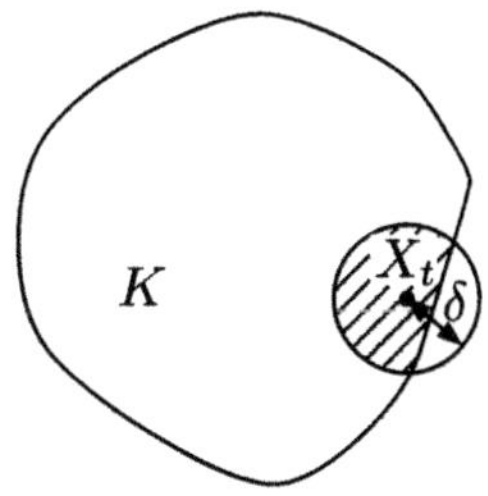

Given a point $X_t \in K$, which is the position of the random walk at time t, we choose X_{t+1} uniformly at random from $B(X_t, \delta) \cap K$, where $B(x, r)$ denotes the ball or radius r centred at x, and δ is a small appropriately chosen constant.[2] We will show that this Markov chain has a stationary distribution that is nearly uniform over K, and that its mixing time is polynomial in the dimension n, provided the step size δ is chosen judiciously, and that K satisfies certain reasonable conditions. The stochastic process (X_t) is Markovian — the distribution of X_{t+1} depends only on X_t and not on the prior history $(X_0, \ldots, X_{t-1})$ — but unlike the Markov chains so far encountered has infinite, even uncountable state space. We therefore pause to look briefly into the basic theory of Markov chains on $\mathbb{R}^n$.

6.1 A few remarks on Markov chains with continuous state space

Our object of study in this chapter is an MC whose state space, namely K, is a subset of $\mathbb{R}^n$. We cannot usefully speak directly of the probability of making a transition from $x \in K$ to $y \in K$, since this probability is generally 0. The solution is to speak instead of the probability $P(x, A) := \Pr[X_1 \in A \mid X_0 = x]$ of being in a (measurable) set $A \subseteq K$ at time 1 conditioned on being at x at time 0. The t step transition probabilities can then be defined inductively by $P^1 := P$ and

$$P^t(x, A) := \int_K P^{t-1}(x, dy)\, P(y, A) \tag{6.1}$$

for $t > 1$. In the case of the ball walk,

$$P(x, A) = \frac{\mathrm{vol}_n(B(x, \delta) \cap A)}{\mathrm{vol}_n(B(x, \delta) \cap K)},$$

for any (measurable) $A \subseteq K$, and

$$P(x, dy) = \frac{dy}{\mathrm{vol}_n(B(x, \delta) \cap K)}, \tag{6.2}$$

[2]What is described here is a "heat-bath" version of the ball wall, which has been termed the "speedy walk" in the literature. There is also a slower "Metropolis" version that we shall encounter presently.

provided $y \in B(x,\delta) \cap K$.

A MC with continuous state space may have one or more invariant measures μ, which by analogy with the finite case means that μ satisfies

$$\mu(A) = \int_K P(x,A)\,\mu(dx),$$

for all measurable sets $A \subseteq K$. As in the finite case, the MC may converge to a unique invariant measure μ in the sense that $P^t(x,A) \to \mu(A)$ as $t \to \infty$ for all $x \in K$ and all measurable $A \subseteq K$.

For compactness, we shall sometimes drop explicit reference to the variable of integration in situations where no ambiguity arises, and write, e.g., $\int_K f\,d\mu$ in place of $\int_K f(x)\,\mu(dx)$.

6.2 Invariant measure of the ball walk

If we were to choose δ, the step-size of the ball walk, to be greater than the diameter $D := \sup\{\|x-y\| : x,y \in K\}$ of K, then the the ball walk would converge in one step to the uniform measure on K. (For convenience, we'll drop the subscript in the Euclidean norm $\|\cdot\|_2$.) There must be a catch! A moment's reflection reveals that the problem is one of implementability: to perform one step of the ball walk when $\delta \geq D$ we must sample a point uniformly at random from K, which is exactly the problem we set ourselves at the outset. However, provided we choose δ small enough, specifically so the ratio $\mathrm{vol}_n\big(B(X_t,\delta)\cap K\big)/\mathrm{vol}_n B(X_t,\delta)$ is not too small, we may obtain a random sample from $B(X_t,\delta) \cap K$ by repeatedly sampling from $B(X_t,\delta)$ until we obtain a point in $B(X_t,\delta) \cap K$. This is the so-called "rejection sampling" method, which is efficient provided that the probability of a successful trial is not too small.

This foregoing observation leads us to introduce a "Metropolis" version of the ball walk (which should be compared with the heat-bath version specified earlier): select a point y u.a.r. from $B(X_t,\delta)$; if $y \in K$ then set $X_{t+1} \leftarrow y$, else set $X_{t+1} \leftarrow X_t$. The Metropolis version of the ball walk has the advantage of implementability over the heat-bath version. However, it has the disadvantage that it can get stuck in sharp corners. Consider what would happen, for example, if the Metropolis walk ended up very close (in relation to the step size δ) to the corner of an n-dimensional cube. To make progress, the point y would have to move in the correct direction in each of the coordinate axes, an event that occurs with probability close to 2^{-n}. So the Metropolis walk cannot be rapidly mixing in the usual sense. We could try to loosen the definition of mixing time by somehow excluding sharp corners as possible initial states, and excluding them also from the metric employed to measure distance from stationarity. But it is cleaner to argue about the mixing time of the heat-bath version of the ball walk, and then separately argue about the relationship of the heat-bath and Metropolis walks.

The primary aim of this chapter is to convey the key ideas underlying the analysis of the ball walk, and not to obtain the most general theorems. We therefore simplify our analysis by imposing a "curvature condition" on K that rules out sharp corners. This condition radically simplifies certain technical aspects of the proof, while leaving intact all the main insights. One immediate effect of this simplification is that the Metropolis walk becomes only a constant factor slower than the heat-bath walk, so we have an easy job relating the two. Towards the end of the section, we shall review the proof and see what extra work needs to be done to eliminate the curvature condition. Provided we are prepared to accept a bound on mixing time that is wrong by a factor of n, the curvature condition may be dropped with little effort. Obtaining the correct mixing time in the absence of the curvature condition requires an analysis of substantial additional technical complexity, but requiring no further significant insights. This improvement will therefore be sketched only.

In the light of the preceeding discussion, we cannot expect the mixing time of the Metropolis version of the ball walk to be short if K is very long and thin. The small "width" of K would dictate a small δ, but then very many steps would be required to get from one end of K to the other. In the full strength version of the bound on mixing time of the ball walk, this issue is resolved by expressing the mixing time in terms of some measure of the "aspect ratio" of K. More precisely, it is supposed that K contains the unit ball $B(0,1)$ centred at the origin, and then the mixing time is expressed as a function of the diameter of K.[3] In fact, as already indicated, we simplify our presentation by making a stronger assumption, namely that the curvature of K should not be too large. We embody this simplifying assumption in the *curvature condition*:

$$\begin{aligned}&\text{For all points } x \in K \text{ there is some point } y \in K\\ &\text{such that } x \in B(y,1) \text{ and } B(y,1) \subseteq K.\end{aligned} \tag{6.3}$$

By definition, all balls will be closed. Note that the curvature assumption is much stronger that the "official" one, which merely asserts that $B(0,1) \subseteq K$ and, in particular, rules out the interesting case of K a polytope. For the main body of this chapter, and until further notice, "ball walk" will implicitly mean the heat-bath version, and the curvature condition will be assumed.

Remark 6.1. What if we *are* presented with a body that is "thin"? It turns out that it is always possible to apply a linear transformation to K to yield a new convex body which contains a unit ball and whose diameter is quite reasonable. But this is another long story, and we do not embark on it here. Refer to Kannan, Lovász and Simonovits [39].

The stationary measure of the ball walk — we shall see presently that the ball walk is ergodic — is not uniform over K, but is close to uniform provided the

[3]Note, as a by-product, we know that K contains the origin, so we have a suitable starting point for the random walk.

step size δ is not too large. To describe the stationary measure, we introduce a function $\ell : K \to \mathbb{R}$ (called *local conductance* by Lovász and Simonovits) defined as

$$\ell(x) := \frac{\mathrm{vol}_n(B(x,\delta) \cap K)}{\mathrm{vol}_n B(x,\delta)}, \tag{6.4}$$

which may be interpreted as the probability of staying in K when choosing a random point in a δ-ball around x. Note that $\ell(x)^{-1}$ is the expected number of repetitions of this trial in order produce a point lying in $B(x,\delta) \cap K$ using rejection sampling. We want to normalise $\ell(x)$ in order to get a density which will turn out to be the density of the stationary measure of the ball walk:

$$\mu(A) := \frac{\int_A \ell(x)\,dx}{L} \quad \text{where} \quad L = \int_K \ell(x)\,dx. \tag{6.5}$$

Our first task is to verify that μ is an invariant measure for the ball walk. That it is unique follows as a weak consequence of our rapid mixing proof.

Lemma 6.2. *If X_0 has distribution μ, then X_1 does also.*

Proof. Let μ_1 denote the distribution of X_1. Then

$$\begin{aligned}
\mu_1(A) = \int_A \mu_1(dy) &= \int_A \int_K P(x,dy)\,\mu(dx) \\
&= \int_A dy \int_{B(y,\delta)\cap K} \frac{\mu(dx)}{\mathrm{vol}_n(B(x,\delta)\cap K)} && \text{by (6.2)} \\
&= \frac{1}{L} \int_A dy \int_{B(y,\delta)\cap K} \frac{\ell(x)\,dx}{\mathrm{vol}_n(B(x,\delta)\cap K)} && \text{by (6.5)} \\
&= \frac{1}{L} \int_A dy \int_{B(y,\delta)\cap K} \frac{dx}{\mathrm{vol}_n B(x,\delta)} && \text{by (6.4)} \\
&= \frac{1}{L} \int_A \ell(y)\,dy = \mu(A) && \text{by (6.4, 6.5).}
\end{aligned}$$

Hence μ is an invariant measure for the ball walk. □

Exercise 6.3. Show that the uniform distribution on K is an invariant measure for the *Metropolis* version of the ball walk.

It is clear that the distribution μ is not uniform over K, but for a suitable choice of δ it is close to it.

Lemma 6.4. *Assume the curvature condition (6.3), and suppose that $\delta \le c_1/\sqrt{n}$ (where c_1 is a dimension-independent constant). Then $0.4 \le \ell(x) \le 1$ for all $x \in K$.*

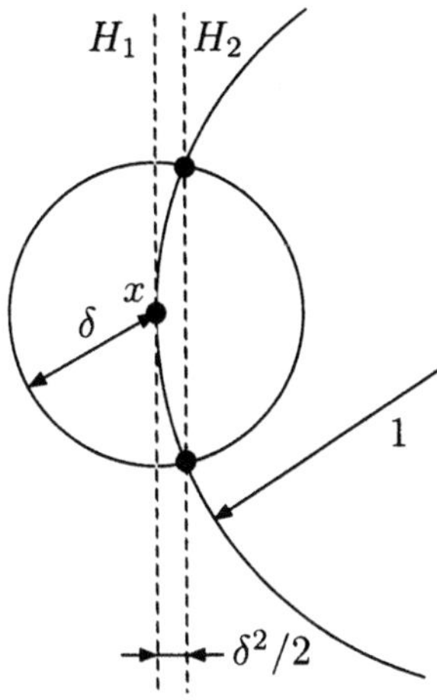

Figure 6.3: Bounding the volume of intersection

Proof. The upper bound on $\ell(x)$ is trivial from the definition of ℓ. For the lower bound we need an argument.

Recall that we assume that every $x \in K$ lies in a 1-ball $B(y,1) \subseteq K$. The inequality above will follow from

$$\frac{\text{vol}_n(B(x,\delta) \cap B(y,1))}{\text{vol}_n B(x,\delta)} \geq 0.4.$$

It is enough to show the relation for a point x on the boundary of $B(y,1)$. Consider the tangent plane H_1 to $B(y,1)$ through x and its parallel plane H_2 through the intersection of the boundaries of the two balls. (Refer to Figure 6.3.) Orient them such that their positive side H_i^+ $(i = 1,2)$ contains the point y. Notice that

$$B(x,\delta) \cap H_2^+ \subset B(y,1)$$

(δ is assumed to be smaller than 1). Therefore it is enough to show that the set $B(x,\delta) \cap H_2^+$ has volume at least $0.4\,\text{vol}_n B(x,\delta)$. We will do this by showing that $B(x,\delta) \cap H_2^- \cap H_1^+$ has very small volume, i.e., at most a 0.1 fraction of the volume of $B(x,\delta)$. The set in question is contained in the cylinder with ground face $B(x,\delta) \cap H_1$ (which is an $(n-1)$-dimensional ball with radius δ) whose height is the distance apart of H_1 and H_2. A simple computation reveals that this distance is exactly $\delta^2/2$. From the volume formula of balls of dimensions $n-1$ and n, and Stirling's approximation for the Γ-function, we obtain the following relation

$$\frac{\text{vol}_{n-1}(B(x,\delta) \cap H_1)}{\text{vol}_n B(x,\delta)} \leq \frac{c\sqrt{n}}{\delta},$$

for some universal constant c. Hence the volume of the cylinder is at most a $\frac{1}{2}c\delta\sqrt{n}$ fraction of the volume of $B(x,\delta)$. Setting $c_1 = 1/5c$ gives the desired bound. □

What this lemma also says is that we can implement one transition of the ball walk efficiently: going from a point $x \in K$ to a random point in $B(x, \delta)$ we have a probability of at least 0.4 of ending up in K immediately; in other words, the Metropolis version of the ball walk is only a factor 2.5 slower than the heat-bath version.

6.3 Mixing rate of the ball walk

We will show now that the ball walk mixes rapidly. The next lemma is a powerful weapon and forms the basis of one of our standard techniques.

Lemma 6.5. *Let f be a measurable function over a measurable set S. Partition S into measurable sets $S_0, \dots, S_{m-1}$. Then*

$$\int_S f^2 \, d\mu = \sum_{i=0}^{m-1} \int_{S_i} (f - \bar{f}_i)^2 \, d\mu + \sum_{i=0}^{m-1} \mu(S_i) \bar{f}_i^2, \tag{6.6}$$

where

$$\bar{f}_i := \frac{1}{\mu(S_i)} \int_{S_i} f \, d\mu.$$

Remark 6.6. Suppose that $\mathrm{E}_\mu f := \int_K f \, d\mu = 0$. Then on the l.h.s. of the equality we have simply $\mathrm{Var}_\mu f$. The two terms on the r.h.s. of the equality may be interpreted as (i) the sum of the variances of f *within* each of the regions S_i, and (ii) the variance of f *between* the regions, respectively.

Proof of Lemma 6.5.

$$\begin{aligned}
\int_{S_i} (f - \bar{f}_i)^2 \, d\mu + \mu(S_i) \bar{f}_i^2 &= \int_{S_i} f^2 \, d\mu + \int_{S_i} \bar{f}_i^2 \, d\mu - 2 \int_{S_i} \bar{f}_i f \, d\mu + \mu(S_i) \bar{f}_i^2 \\
&= \int_{S_i} f^2 \, d\mu + \mu(S_i) \bar{f}_i^2 - 2\mu(S_i) \bar{f}_i^2 + \mu(S_i) \bar{f}_i^2 \\
&= \int_{S_i} f^2 \, d\mu.
\end{aligned}$$

□

As in the analysis of the matchings MC, our approach to bounding the mixing time involves taking a (measurable) test function $f : K \to \mathbb{R}$ (with $\mathrm{E} f = 0$ for convenience) and examining how the variance of f decays as a result of the averaging effect of the ball-wall. To this end, introduce a function $h : K \to \mathbb{R}$ given by

$$\begin{aligned}
h(x) &:= \frac{1}{2} \int_K P(x, dy) \, (f(x) - f(y))^2 \\
&= \frac{1}{2 \, \mathrm{vol}_n(B(x, \delta) \cap K)} \int_{B(x,\delta) \cap K} (f(x) - f(y))^2 \, dy,
\end{aligned} \tag{6.7}$$

and define

$$\operatorname{Var}_\mu f := \int_K f^2 \, d\mu \quad \text{and} \quad \mathcal{E}_\mu(f,f) := \int_K h \, d\mu;$$

these are the now-familiar variance (global variation of f over K) and Dirichlet form (local variation of f at the scale of the step size δ of the ball walk). As with the matching MC, the key to the analysis of the ball walk lies in obtaining a sharp Poincaré inequality linking $\operatorname{Var}_\mu f$ and $\mathcal{E}_\mu(f,f)$. Our eventual goal is to show:

Theorem 6.7 (Poincaré inequality). *Let $K \subset \mathbb{R}^n$ be a convex body of diameter D satisfying the curvature condition (6.3), and suppose that δ is as in Lemma 6.4. For any (measurable) function $f : K \to \mathbb{R}$,*

$$\mathcal{E}_\mu(f,f) \geq \lambda \operatorname{Var}_\mu f \tag{6.8}$$

where

$$\lambda := \frac{c_2 \delta^2}{D^2 n}$$

for some universal constant c_2.

We apply the technique by Mihail (as we did with matchings in §5.2) and obtain from λ a bound on mixing time. As before, we deal with periodicity by considering either a continuised or lazy walk.

Corollary 6.8. *For any $\varepsilon > 0$ let $\tau(\varepsilon)$ denote the time at which the ball walk (in either its continuised or lazy variants) reaches within total variation distance ε of the stationary distribution μ. Then, under the curvature condition (6.3),*

$$\tau(\varepsilon) \leq O\left(\lambda^{-1}\left(\log \varepsilon^{-1} + i(\mu_0)\right)\right),$$

where λ is as in Theorem 6.7 and $i(\mu_0)$ expresses the dependence on the initial distribution μ_0.

Remark 6.9. The expression $i(\mu_0)$ is closely related to the term $\ln \pi(x_0)^{-1}$ familiar from the discrete case. But if we now start from a fixed point (in other words our initial distribution μ_0 is a single point mass at $x_0 \in K$) no meaning can be attached to $\ln \pi(x_0)^{-1}$. To escape from this, imagine that we start at time -1 from a point x_0 such that $B(x_0,\delta) \subseteq K$, and consider the situation at time 0. Thus the initial distribution μ_0 is uniform over some ball of radius δ. In this case, we may take $i(\mu_0) = n \ln(D/2\delta)$.

Exercise 6.10. Verify Corollary 6.8. Doing this essentially involves translating Theorem 5.6 to the setting of continuous state space. In case you skip this exercise, a full derivation may be found in §6.8.

At an intuitive level, Theorem 6.7 seems to be close to the truth. With a step size of δ, the distance travelled parallel to any axis fixed in advance (in particular, one parallel to a diameter of K) is of order $\delta/\sqrt{n}$. The time taken for the walk to "diffuse" along a diameter is the square of the ratio of D to the typical distance moved along the diameter in one step, namely $(D\sqrt{n}/\delta)^2$, which is of order λ^{-1}. To minimise mixing time we clearly wish to take δ as small as possible consistent with implementability, which by Lemma 6.4 is of order $n^{-1/2}$. With that step size, the Poincaré constant scales as $(nD)^{-2}$.

The next section is devoted to the proof of what is essentially the main result of this chapter.

6.4 Proof of the Poincaré inequality (Theorem 6.7)

Assume the converse to (6.8), namely that there exists a function $f : K \to \mathbb{R}$ with

$$\mathcal{E}_\mu(f, f) < \lambda \operatorname{Var}_\mu f; \tag{6.9}$$

informally, f sustains high global variation simultaneously with low local variation.

We will define smaller and smaller *violating sets* S such that the ratio

$$\int_S h\, d\mu \bigg/ \int_S (f - \bar{f})^2\, d\mu \tag{6.10}$$

is small, where $\bar{f} = \int_S f\, d\mu$. Our starting point is of course $S = K$, where we know that this ratio is less than λ. Eventually, S will be small even with respect to δ. Then the function f will have to be almost constant in S since the local variation (as measured by the numerator) is small; however the global variation (as measured by the denominator) is large. Here we reach a contradiction. This in outline is our proof.

First we will shrink the violating set to a set K_1 which is very small in all but one dimension, a so-called "needle-like" body. It transpires that we can do this while keeping ratio (6.10) bounded throughout by λ. It is only when we attempt to shrink along the final dimension that we have to give something away. Before embarking on the process of shrinking K to a needle-like body, we need a pair of geometrical lemmas, whose proofs we defer to §6.5.

Lemma 6.11. *Let R be a convex set in $\mathbb{R}^2$. There is a point $x \in R$ such that every line through x partitions R into pieces of area at least $\frac{1}{3}$ of the area of R.*

Remark 6.12. The bound $\frac{1}{3}$ can in fact be replaced by $\frac{4}{9}$, which is tight as can been seen by considering an equilateral triangle; see Egglestone [26, §6.4]. However, any strictly positive bound is adequate for our purposes.

The *width* of a convex set R in $\mathbb{R}^2$ is the minimum, over all pairs of parallel supporting lines sandwiching R, of the distance between those lines.[4]

[4] In some sense, width it is the opposite of diameter, which may be defined as the maximum such distance. This was not how we defined diameter in §6.2, but the two definitions are equivalent.

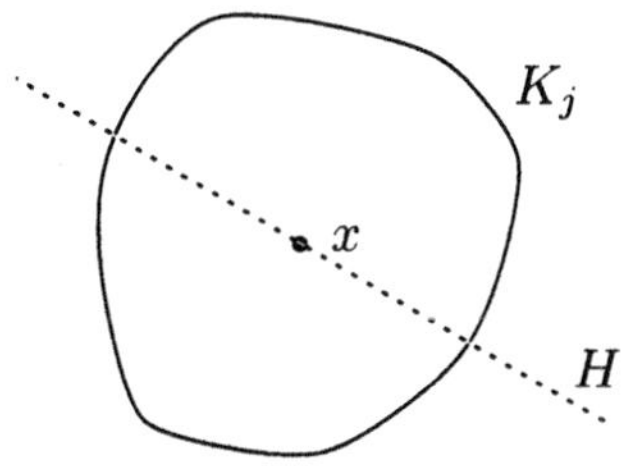

Figure 6.4: The expectation of f is zero on both $K_j \cap H^+$ and $K_j \cap H^-$.

Lemma 6.13. *Let R be a convex set in $\mathbb{R}^2$ of area A. Then the width of R is at most $\sqrt{2A}$.*

Remark 6.14. Again, the bound is not the best possible, but is adequate for our purposes. The extremal set (i.e., the one of given area that maximises width) is again an equilateral triangle.

To resume: With the aim of establishing a contradiction we are assuming the existence of a function $f : K \to \mathbb{R}$ satisfying (6.9). We may further assume (by adding an appropriate constant function to f) that $\mathrm{E}_\mu f = 0$. This additional assumption will be convenient on the first leg of our journey towards the contradiction.

Claim 6.15. *Assume $f : K \to \mathbb{R}$ satisfies inequality (6.9), and $\mathrm{E}_\mu f = 0$. Then, for every $\varepsilon > 0$, there is a convex subset $K_1 \subseteq K$ satisfying*

$$\int_{K_1} h\,d\mu < \lambda \int_{K_1} f^2\,d\mu \quad \textit{as well as} \quad \int_{K_1} f\,d\mu = 0,$$

and such that K_1 lies in the box $[0,D] \times [0,\varepsilon]^{n-1}$ in some Cartesian coordinate system.

Proof. Suppose, for some $j \geq 2$, that K_j is a violating set which lies in $[0,D]^j \times [0,\varepsilon]^{n-j}$, and that $\int_{K_j} f\,d\mu = 0$; i.e., we have already shrunk our violating set down on $n-j$ coordinates. (The base case $K_n = K$ is of course covered by (6.9).) To shrink along a further coordinate we use a beautiful divide-and-conquer argument due to Payne and Weinberger: see Bandle [4, Thm 3.24].

Let R be the projection of K_j onto the first two (i.e., "fat") axes. Let x be a point satisfying the conditions of Lemma 6.11. Consider all $(n-1)$-dimensional planes through x whose normals lie in the 2-dimensional plane spanned by the first two axes. These planes project to lines through x in the plane of R. Among these planes there is at least one, say H, such that

$$\int_{K_j \cap H^+} f\,d\mu = \int_{K_j \cap H^-} f\,d\mu = 0.$$

To see this, choose any $(n-1)$-dimensional plane G through x whose normal lies within the plane of R. If G does not already have the desired property, then, since $\int_{K_j \cap G^+} f \, d\mu + \int_{K_j \cap G^-} f \, d\mu = 0$, one integral or the other has to be positive. By rotating G about x by an angle of π, the signs exchange. So by continuity and the mean value theorem we have to have hit the sought-for H at some point.

It is easy to convince oneself that K_j intersected with one side of H (i.e., either $K_j \cap H^+$ or $K_j \cap H^-$) is also a violating set, in the sense that the ratio (6.10) is bounded by λ when $S = K_j \cap H^+$ (or $S = K_j \cap H^-$, as appropriate). Now iterate this procedure. By Lemma 6.11, the area of the projection R of the convex body drops by a constant factor at each iteration, and must eventually drop below $\frac{1}{2}\varepsilon^2$. At this point the width of R, by Lemma 6.13, is at most ε. Then, rotating the fat axes as appropriate, the projection of the convex body onto (say) the first of these axes is a line segment of length at most ε. The convex set now has exactly the properties we require of the set K_{j-1}, i.e., the same properties as K_j, but with $j-1$ replacing j. Hence by induction we can find our set K_1. □

The above line of argument requires at least two fat dimensions in order to provide enough freedom in selecting the plane H. We need a new approach in order to shrink the needle-line set along the remaining fat dimension.

Claim 6.16. *Let K_1 and f be as in the conclusion of Claim 6.15, δ be as in Lemma 6.4, and let $\eta := c_3\delta/\sqrt{n}$ where $c_3 > 0$ is any constant. Then, under the curvature condition (6.3), there is a convex subset $K_0 \subseteq K_1$ satisfying*

$$\int_{K_0} h \, d\mu < \frac{1}{10} \int_{K_0} (f - \bar{f})^2 \, d\mu \tag{6.11}$$

where

$$\bar{f} = \frac{1}{\mu(K_0)} \int_{K_0} f \, d\mu, \tag{6.12}$$

and such that K_0 lies in the box $[-\eta, \eta] \times [0, \varepsilon]^{n-1}$ in some Cartesian coordinate system.

Remark 6.17. We will choose the constant c_3 later; in order to obtain an eventual contradiction, it will need to be small enough. The choice of c_3 will then determine the universal constant c_2 of Theorem 6.7: the smaller c_3, the smaller c_2.

Our strategy for proving Claim 6.16 is to chop K_1 into short sections and show that at least one of these sections (or perhaps the union of two adjacent ones) satisfies the inequality (6.11). (Refer to Figure 6.5.) Before embarking on the proof proper, we need another geometric lemma, which is a consequence of the Brunn-Minkowski Theorem; the proof is again deferred to §6.5.

Lemma 6.18. *Let convex body K_1 be partitioned into m pieces $S_0 \dots S_{m-1}$ of equal width by planes orthogonal to a fixed axis. Then the sequence*

$$\frac{1}{\mathrm{vol}_n S_0}, \frac{1}{\mathrm{vol}_n S_1}, \dots, \frac{1}{\mathrm{vol}_n S_{m-1}}$$

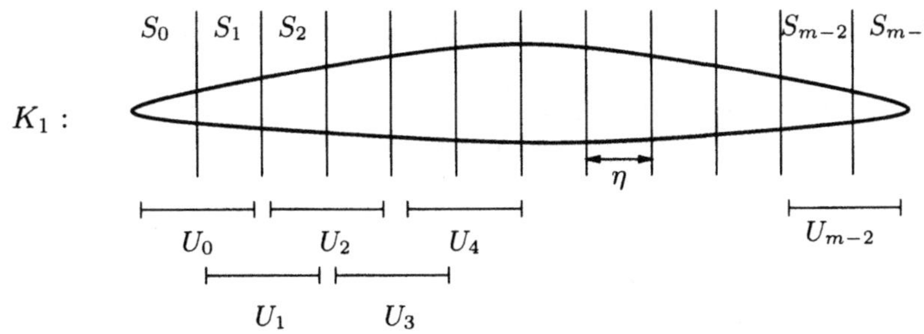

Figure 6.5: Partitioning of K_1

is convex.

We are ready to resume the chopping argument.

Proof of Claim 6.16. Let convex body K_1 be partitioned into m pieces by planes orthogonal to the fat axis, as specified in Lemma 6.18, so that each piece S_i has width $\eta = c_3\delta/\sqrt{n}$. Additionally, define $U_i := S_i \cup S_{i+1}$ for $i = 0, 1, \ldots, m-2$. Note that $m = O(D\sqrt{n}/\delta)$. Using Lemma 6.5, we find

$$\int_{K_1} f^2 \, d\mu = \underbrace{\sum_{i=0}^{m-1} \int_{S_i} (f - \bar{f}_i)^2 \, d\mu}_{A} + \underbrace{\sum_{i=0}^{m-1} \mu(S_i) \bar{f}_i^2}_{B}, \tag{6.13}$$

where for convenience we define

$$\bar{f}_i := \frac{1}{\mu(S_i)} \int_{S_i} f \, d\mu.$$

In the case that sum A is greater or equal to sum B, we readily find a piece S_i that serves as a violating set. We start with

$$\begin{aligned} \sum_{i=0}^{m-1} \int_{S_i} h \, d\mu &= \int_{K_1} h \, d\mu && (6.14) \\ &< \lambda \int_{K_1} f^2 \, d\mu && \text{by assumption} \\ &\le 2\lambda \sum_{i=0}^{m-1} \int_{S_i} (f - \bar{f}_i)^2 \, d\mu && \text{by (6.13) and } A \ge B. \quad (6.15) \end{aligned}$$

Comparing sums (6.14) and (6.15) we see there must be an S_i such that

$$\int_{S_i} h \, d\mu \le 2\lambda \int_{S_i} (f - \bar{f}_i)^2 \, d\mu.$$

Setting $K_0 = S_i$ satisfies the conclusion of the claim with plenty to spare. (Note in this context that $\lambda = O(n^{-2})$.)

The case $B > A$ is a little more difficult. Using the alternative expression for variance which we have seen before, and recalling that the expectation of f with respect to μ on K_1 is 0, we have

$$\begin{aligned} \mu(K_1)\int_{K_1} f^2\,d\mu &< 2\,\mu(K_1)\sum_{i=0}^{m-1}\mu(S_i)\bar{f}_i^2 && \text{since } B > A \\ &= 2\sum_{0\le i<j<m}\mu(S_i)\mu(S_j)(\bar{f}_i - \bar{f}_j)^2 && \text{using (5.5).} \end{aligned} \tag{6.16}$$

Our aim is to replace the r.h.s. of (6.16) by a sum with similar terms, but restricted to *adjacent* pairs i, j. This will enable us to finish with an argument similar to the $A \ge B$ case.

For convenience, we introduce the abbreviation $w_i = \mu(S_i)$, and set

$$a_{i,j} := w_i w_j \sum_{k=i}^{j-1}\frac{w_k + w_{k+1}}{w_k w_{k+1}} \le 2 w_i w_j \sum_{k=i}^{j}\frac{1}{w_k}. \tag{6.17}$$

Inequality (6.16) may be massaged as follows:

$$\begin{aligned} \mu(K_1)\int_{K_1} f^2\,d\mu &< 2\sum_{i<j} w_i w_j(\bar{f}_i - \bar{f}_j)^2 \\ &= 2\sum_{i<j} w_i w_j\left[\sum_{k=i}^{j-1}(\bar{f}_k - \bar{f}_{k+1})\right]^2 \\ &= 2\sum_{i<j} w_i w_j\left[\sum_{k=i}^{j-1}\sqrt{\frac{w_k + w_{k+1}}{w_k w_{k+1}}}\cdot\sqrt{\frac{w_k w_{k+1}}{w_k + w_{k+1}}}\,(\bar{f}_k - \bar{f}_{k+1})\right]^2 \\ &\le 2\sum_{i<j} a_{i,j}\sum_{k=i}^{j-1}\frac{w_k w_{k+1}}{w_k + w_{k+1}}(\bar{f}_k - \bar{f}_{k+1})^2, \end{aligned} \tag{6.18}$$

where the final inequality is Cauchy-Schwarz combined with (6.17). Define $\hat{f}_k$ to be the expectation of f over $U_k = S_k \cup S_{k+1}$:

$$\hat{f}_k := \frac{1}{\mu(U_k)}\int_{U_k} f\,d\mu = \frac{w_k\bar{f}_k + w_{k+1}\bar{f}_{k+1}}{w_k + w_{k+1}}.$$

Then, by Lemma 6.5,

$$\begin{aligned} \frac{w_k w_{k+1}}{w_k + w_{k+1}}(\bar{f}_k - \bar{f}_{k+1})^2 &= w_k(\bar{f}_k - \hat{f}_k)^2 + w_{k+1}(\bar{f}_{k+1} - \hat{f}_k)^2 \\ &\le \int_{U_k}(f - \hat{f}_k)^2\,d\mu \end{aligned} \tag{6.19}$$

(The first line may be viewed as the special case $|\Omega| = 2$ of (5.5), or may be verified by elementary algebraic manipulation. Inequality (6.19) comes from Lemma 6.5, noting that the first sum on the r.h.s. of (6.6) is clearly positive.) Applying bound (6.19) to the terms in (6.18) yields

$$\mu(K_1)\int_{K_1} f^2\,d\mu < 2\sum_{i<j} a_{i,j}\sum_{k=i}^{j-1}\int_{U_k}(f-\hat{f}_k)^2\,d\mu. \tag{6.20}$$

Taking stock momentarily: inequality (6.20) appears to be telling us that if the variance of f is large on K_1 then it must be large on *some* U_k; but there is still some work to be done on the way to quantifying this effect.

Recall that

$$w_i = \mu(S_i) = L^{-1}\int_{S_i}\ell(x)\,dx,$$

where $L = \int_K \ell(x)\,dx$. Thus, by Lemma 6.4,

$$0.4\,L^{-1}\,\mathrm{vol}_n\,S_i \le w_i \le L^{-1}\,\mathrm{vol}_n\,S_i, \tag{6.21}$$

leading to the following upper bound on $a_{i,j}$:

$$\begin{aligned} a_{i,j} &\le 2\,w_i w_j \sum_{k=i}^{j}\frac{L}{0.4\,\mathrm{vol}_n\,S_k} && \text{by (6.17) and (6.21)}\\ &\le 2.5\,w_i w_j L\,(j-i+1)\Big(\frac{1}{\mathrm{vol}_n\,S_i}+\frac{1}{\mathrm{vol}_n\,S_j}\Big) && \text{by Lemma 6.18}\\ &\le 2.5(j-i+1)(w_i+w_j) && \text{by (6.21).} \end{aligned} \tag{6.22}$$

Since $j-i+1$ never exceeds m, we have the following crude bound on the sum of the $a_{i,j}$:

$$\begin{aligned} \sum_{i<j} a_{i,j} &\le 2.5\sum_{i<j}(j-i+1)(w_i+w_j)\\ &\le 2.5\,m\sum_{i<j}(w_i+w_j)\\ &\le 2.5\,m^2\sum_i w_i && (6.23)\\ &= 2.5\,m^2\mu(K_1). && (6.24) \end{aligned}$$

To see (6.23), fix attention on a particular index k and note that w_k occurs exactly $m-1$ times in the double sum.

Returning now to (6.20),

$$\begin{aligned}\mu(K_1)\int_{K_1} f^2\,d\mu &< 2\sum_{i<j} a_{i,j}\sum_{k=i}^{j-1}\int_{U_k}(f-\hat f_k)^2\,d\mu\\ &\le 2\sum_{i<j} a_{i,j}\sum_{k=0}^{m-2}\int_{U_k}(f-\hat f_k)^2\,d\mu\\ &\le 5m^2\mu(K_1)\sum_{k=0}^{m-2}\int_{U_k}(f-\hat f_k)^2\,d\mu \qquad \text{by (6.24)},\end{aligned}$$

from which

$$\int_{K_1} f^2\,d\mu \le 5m^2\sum_{k=0}^{m-2}\int_{U_k}(f-\hat f_k)^2\,d\mu. \tag{6.25}$$

Inequality (6.25) is the one we sought, expressing the fact that if the variance of f is large on the whole of K_1 then it must be fairly large on some piece U_k. Proceeding now by analogy with the $A \le B$ case, using (6.25) and the conclusion of Claim 6.15,

$$\sum_{k=0}^{m-2}\int_{U_k} h\,d\mu \le 2\int_{K_1} h\,d\mu < 2\lambda\int_{K_1} f^2\,d\mu \le 10m^2\lambda\sum_{k=0}^{m-2}\int_{U_k}(f-\hat f_k)^2\,d\mu.$$

Therefore there must exist a k such that

$$\int_{U_k} h\,d\mu < 10m^2\lambda\int_{U_k}(f-\hat f_k)^2\,d\mu. \tag{6.26}$$

By setting c_2 sufficiently small, specifically $c_2 < c_3^2/100$, we obtain

$$10m^2\lambda = 10\left(\frac{D\sqrt{n}}{c_3\delta}\right)^2\frac{c_2\delta^2}{D^2n} < \frac{1}{10}.$$

Setting $K_0 := U_k$, we recognise (6.26) as the inequality promised in the statement of the claim. This concludes the case $B > A$ and hence the proof. □

We pick up the proof of Theorem 6.7. At the outset we assumed, with a view to obtaining a contradiction, the converse of (6.8). Now, from Claims 6.15 and 6.16, we deduce the existence of a convex set $K_0 \subset K$ satisfying inequality (6.11) such that K_0 is contained in a prism of height 2η whose cross section is an $(n-1)$-dimensional cube of side ε. We are close to obtaining the desired contradiction.

Let C be the centre axis of the prism, and let z_1 and z_2 be the points at which C intersects the end facets of the prism. (Refer to Figure 6.6.) Let $\delta' := \delta - \varepsilon\sqrt{n}$, and choose ε sufficiently small that

$$\mathrm{vol}_n\, B(0,\delta') \ge 0.9\,\mathrm{vol}_n\, B(0,\delta). \tag{6.27}$$

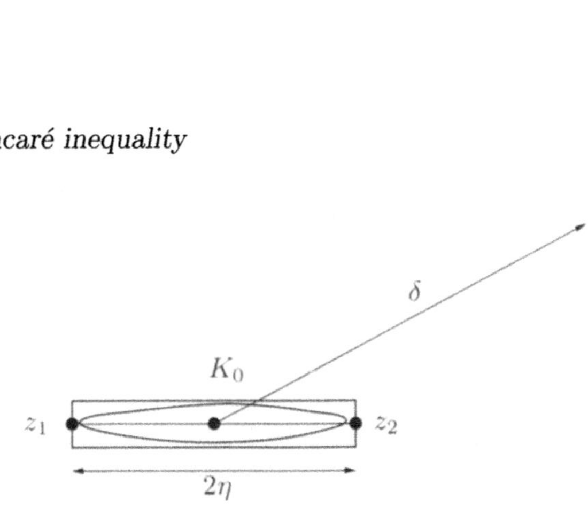

Figure 6.6: "Needle like" body K_0

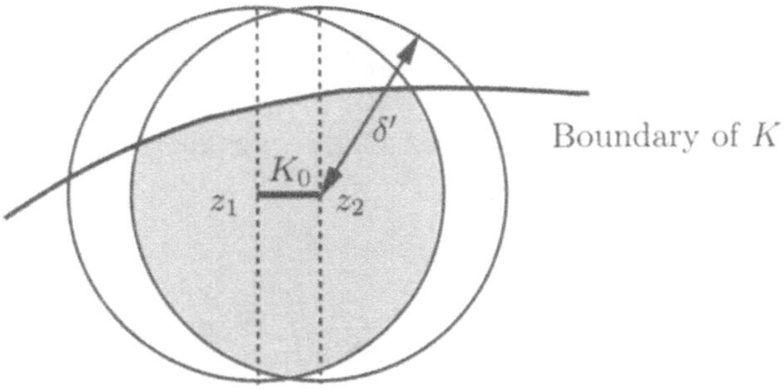

Figure 6.7: Construction of the set I (shown shaded)

(Recall that we are free to choose ε as small as we like.) Set $I := B(z_1, \delta') \cap B(z_2, \delta') \cap K$. (Refer to Figure 6.7.) We shall argue that by choosing c_3 (and hence η) sufficiently small we can ensure

$$\operatorname{vol}_n \big(B(z_1, \delta') \cap B(z_2, \delta')\big) \geq 0.8 \operatorname{vol}_n B(0, \delta), \tag{6.28}$$

and hence

$$\operatorname{vol}_n I = \operatorname{vol}_n \big(B(z_1, \delta') \cap B(z_2, \delta') \cap K\big) \geq 0.2 \operatorname{vol}_n B(0, \delta). \tag{6.29}$$

The calculation supporting (6.28) proceeds exactly as in the proof of Lemma 6.4. Divide $B(z_1, \delta') \cap B(z_2, \delta')$ into two congruent pieces by the plane bisecting the line (z_1, z_2) and orthogonal to it. Each piece can be viewed as a half-ball less a segment that can be contained in a cylinder of height η $(= c_3\delta/\sqrt{n})$ and radius $\delta' \leq \delta$. By setting c_3 small enough — refer to the calculation in the proof of Lemma 6.4 — we may ensure that the volume of this cylinder is less than $0.05 \operatorname{vol}_n B(0, \delta)$. Now, by (6.27), the combined volume of the two half balls is at least $0.9 \operatorname{vol}_n B(0, \delta)$, so after removing the two segments we are still left with a set of volume $0.8\,\mathrm{vol}_n\, B(0,\delta)$, as claimed in (6.28). Inequality (6.29) is now immediate: just observe that the piece of $B(z_1,\delta') \cap B(z_2,\delta')$ that we loose when we intersect with K is contained in $B(z_1,\delta) \setminus K$, which by Lemma 6.4 has volume at most $0.6\,\mathrm{vol}_n\, B(0,\delta)$.

Inequality (6.29) expresses one key property of I, namely that its volume is not too small. The other key property is that every point in I may be reached from any point in K_0 in one step of the ball walk. For by construction,

$$\sup\big\{\|x-y\| : x \in C \text{ and } y \in I\big\} \leq \delta',$$

from which, by the triangle inequality,

$$\sup\big\{\|x-y\| : x \in K_0 \text{ and } y \in I\big\} \leq \delta' + \varepsilon\sqrt{n} = \delta.$$

Since $I \subseteq K$, we may conveniently reformulate this fact as

$$I \subseteq B(x,\delta) \cap K, \quad \text{for all } x \in K_0. \tag{6.30}$$

So,

$$\begin{aligned}
\int_{K_0} h\, d\mu &\geq \frac{1}{2}\int_{K_0} \frac{\mu(dx)}{\mathrm{vol}_n(B(x,\delta)\cap K)} \int_I \big(f(x)-f(y)\big)^2\, dy && \text{by (6.7, 6.30)}\\
&\geq \frac{1}{2\,\mathrm{vol}_n\, B(0,\delta)} \int_{K_0} \mu(dx) \int_I \big(f(x)-f(y)\big)^2\, dy\\
&\geq \frac{1}{2\,\mathrm{vol}_n\, B(0,\delta)} \int_I dy \int_{K_0} \big(f(x)-f(y)\big)^2\, \mu(dx) && \text{(Fubini)}\\
&\geq \frac{1}{2\,\mathrm{vol}_n\, B(0,\delta)} \int_I dy \int_{K_0} (f-\bar{f})^2\, d\mu && (6.31)\\
&\geq \frac{1}{10}\int_{K_0} (f-\bar{f})^2\, d\mu && \text{by (6.29),}
\end{aligned}$$

where $\bar{f}$, as in (6.12), is the μ-expectation of f over K_0. Inequality (6.31) uses a simple fact about variance, namely that $\int_{K_0}(f-c)^2\, d\mu$ is minimised by setting $c = \bar{f}$. But the combined inequality contradicts (6.11). This completes the proof of Theorem 6.7.

6.5 Proofs of the geometric lemmas

In this section we tie up the loose ends by providing proofs for the three geometric lemmas used in the proof of Theorem 6.7.

Proof of Lemma 6.11. The following proof is due to Alan Riddell; I thank him and also Toby Bailey for communicating it to me.

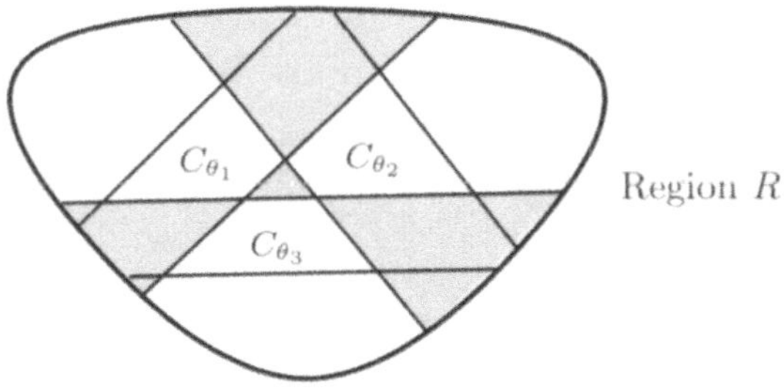

Figure 6.8: A paradoxical subset of R.

Consider all possible partitions of R into three regions of equal area by a pair of parallel lines. (There is one partition corresponding to each orientation for the lines.) Let $\{C_\theta : 0 \le \theta < \pi\}$ be an indexing of the central bands in these partitions, considered as closed sets. Suppose there exist bands C_{θ_1}, C_{θ_2} and C_{θ_3} with no point in common. The set $\mathbb{R}^2 \setminus (C_{\theta_1} \cup C_{\theta_2} \cup C_{\theta_3})$ consists of six unbounded regions and one triangle. Consider the partition of R into seven pieces obtained by extending the edges of the triangle to the boundary of R, and in particular the four pieces shown shaded in Figure 6.8. Each of the shaded pieces other than the central triangle has area at least $\frac{1}{3}\operatorname{vol}_2 R$, since it is the intersection of two regions of R of area $\frac{2}{3}\operatorname{vol}_2 R$. The central triangle itself has positive area. Thus the total shaded area exceeds $\operatorname{vol}_2 R$, a contradiction.

Hence every triple from $\{C_\theta\}$ has a common point and, by Helly's theorem (see Egglestone [26, Thm 17]), the intersection $\bigcap_\theta C_\theta$ of all central bands is non-empty. Any point in this intersection will do as our choice for x. □

Proof of Lemma 6.13. Suppose R is a convex region in $\mathbb{R}^2$ of area A. Let ℓ_1 and ℓ_1' be parallel supporting lines of R, touching R at the points α and α'. We may arrange for lines ℓ_1 and ℓ_1' to be perpendicular to the line segment $[\alpha, \alpha']$, e.g., by choosing $[\alpha, \alpha']$ to be a diameter of R. Now let ℓ_2 and ℓ_2' be supporting lines of R perpendicular to ℓ_1 and ℓ_1', touching R at the points β and β'. The rectangle formed by these supporting lines has area at least w^2, where w is the width of R. It is easy to see that the convex hull of $\{\alpha, \alpha', \beta, \beta'\}$ has area $\frac{1}{2}w^2$. (The fact that $[\alpha, \alpha']$ is parallel with an edge of the rectangle is crucial here.) But the convex hull of $\{\alpha, \alpha', \beta, \beta'\}$ is contained within R. It follows that $A \ge \frac{1}{2}w^2$. □

Proof of Lemma 6.18. For what follows, we abbreviate $\operatorname{vol}_n S_i$ by v_i. In order to prove the lemma, the notation of Minkowski sums is useful: Let A and B be sets of points and λ a real number. A point p is represented by the vector pointing from 0 to p. Then we define the set $A + B$ as the set of points $a + b$ with $a \in A$ and $b \in B$. Furthermore, for a scalar λ, λA is the set of points λa with $a \in A$.

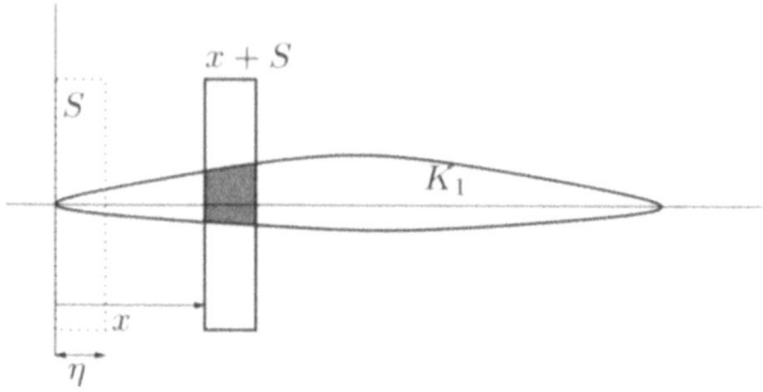

Figure 6.9: Slab S sweeping over K_1

We prove the lemma by showing properties of the function $\text{vol}_n\big((xe_1 + S) \cap K_1\big)$ for $x \in [0, D]$, where S is a "slab" of width η, and e_1 is a unit vector parallel to the fat axis. (The slab is defined as the intersection of two halfspaces orthogonal to the fat axis and distant η apart; assume that the origin is placed at the leftmost point of K_1.) Thus we move the slab S from left to right and observe how the volume of the intersection $K_1 \cap S$ behaves. Note that $v_i := \text{vol}_n S_i = \text{vol}_n\big((i\eta e_1 + S) \cap K_1\big)$. (Refer to Figure 6.9.)

The proof of the lemma relies on a theorem of Brunn and Minkowski (see Egglestone [26, Thm 46]).

Theorem 6.19 (Brunn-Minkowski). *Let K' and K'' be two convex bodies in $\mathbb{R}^n$. Then*

$$\text{vol}_n(K' + K'')^{1/n} \geq \text{vol}_n(K')^{1/n} + \text{vol}_n(K'')^{1/n}.$$

To continue with the proof of Lemma 6.18, observe that

$$(\lambda x + (1-\lambda)y + S) \cap K_1 \supseteq \lambda((x + S) \cap K_1) + (1-\lambda)((y+S) \cap K_1). \tag{6.32}$$

To verify this, assume z is in the set on the right hand side. This means that we can write $z = z' + z''$ with $z' \in \lambda((x+S) \cap K_1)$ and $z'' \in (1-\lambda)((y+S) \cap K_1)$. Therefore, $z' \in \lambda K_1$ and $z'' \in (1-\lambda)K_1$. Thus $z \in K_1$. On the other hand, we have $z' \in \lambda(x+S)$ and $z'' \in (1-\lambda)(y+S)$ which leads to $z \in \lambda x + (1-\lambda)y + S$.

Using the Brunn-Minkowski Theorem in conjunction with (6.32), we find

$$\begin{aligned}
&\text{vol}_n\left[(\lambda x + (1-\lambda)y + S) \cap K_1\right]^{1/n} \\
&\quad \geq \text{vol}_n\left[\lambda((x+S) \cap K_1) + (1-\lambda)((y+K_1) \cap K_1)\right]^{1/n} \\
&\quad \geq \text{vol}_n[\lambda((x+S) \cap K_1)]^{1/n} + \text{vol}_n[(1-\lambda)((y+S) \cap K_1)]^{1/n} \\
&\quad = \lambda \,\text{vol}_n[(x+S) \cap K_1]^{1/n} + (1-\lambda)\,\text{vol}_n[(y+S) \cap K_1]^{1/n}.
\end{aligned}$$

In the last step, we used $\mathrm{vol}_n(\lambda K) = \lambda^n \mathrm{vol}_n K$. As a special case of this inequality, we find that the sequence $(v_i^{1/n})$ is concave:

$$2v_i^{1/n} \geq v_{i-1}^{1/n} + v_{i+1}^{1/n} \tag{6.33}$$

Now it is easily checked that if (a_i) is any concave sequence, and g any monotone non-increasing convex function, then the sequence $(g(a_i))$ is convex. The lemma then follows from (6.33) by setting $a_i = v_i^{1/n}$ and $g(x) = x^{-n}$. □

6.6 Relaxing the curvature condition

What happens if we do not have the curvature condition (6.3)? As we shall see, the question is of some importance, not least because the standard reduction from volume estimation to sampling introduces sharp corners, even if these are absent in the given convex body K. The most obvious consequence of dropping (6.3) is that the expected number of Metropolis steps to simulate a single heat-bath step is no longer bounded by a constant. Worse, as we have argued, the expected number steps may be exponential in n for a worst-case choice for the current point $X_t = x$. The most we can hope for is that, in a typical evolution of the ball walk, we are very unlikely to visit this bad region of K. This turns out indeed to be the case, provided $\delta = O(1/\sqrt{n})$, the body K contains the unit ball $B(0,1)$, and we make a reasonable choice of initial state. See Kannan, Lovász and Simonovits [39].

Remark 6.20. To get a feel for what is going on, imagine the Metropolis ball walk in some n-dimensional polytope K. In order to mix, the walk needs potentially to "see all the boundary" of K, otherwise it cannot gain information about the body. In the case of a polytope this means that we would have to treat the case of coming close to *facets* (i.e., $(n-1)$-dimensional faces) of the polytope. There the random walk can "learn" a lot about the shape of K. But it does not necessarily have to come close to smaller-dimensional faces, where the walk might get stuck for long periods. Not surprisingly, the main technical difficulties then arise from showing that close encounters with low-dimensional faces are rare.

A problem arises, however, before we ever reach the comparison of the heat-bath and Metropolis versions of the ball walks. Specifically, our derivation of the key Poincaré inequality contained in Theorem 6.7 made use of the curvature condition at two points: at inequalities (6.21) and (6.29), both of which rely on Lemma 6.4, and both of which fail in the absence of (6.3).

We may avoid the first of these inequalities entirely, thus removing the curvature condition (6.3) from the statement of Claim 6.16. First we make some observations concerning the local conductance ℓ.

Lemma 6.21. *The local conductance ℓ defined in (6.4) satisfies:*

(i) *$\ell(x)^{1/n}$ is concave over K;*

(ii) $\ln \ell$ *is Lipschitz; specifically* $\left|\ln \ell(x) - \ln \ell(y)\right| \le \frac{n}{\delta}\,\|x-y\|$, *for all* $x, y \in K$.

Proof (sketch). We are in a similar situation to that already encountered in the proof of Lemma 6.18: a convex body — there a slab defined by parallel $(n-1)$-dimensional planes, here a ball of radius δ — is translated in a straight line and its intersection with K studied with the aid of the Brunn-Minkowski Theorem (Theorem 6.19). The proof of part (i) here is analogous.

For part (ii), observe that the definition of the function ℓ, presented in (6.4), makes sense outside its official domain, namely K. Observe also that part (i) continues to hold over the larger region $K + B(0,\delta)$, the Minkowski sum of K and the ball of radius δ. Given $x, y \in K$, let z be the point colinear with x and y, at distance δ from y, and on the opposite side of y to x. Note that $z \in K + B(0,\delta)$. Thus, by part (i),

$$\delta\,\ell(x)^{1/n} + \|x-y\|\,\ell(z)^{1/n} \le (\delta + \|x-y\|)\,\ell(y)^{1/n},$$

and hence

$$\frac{\ell(x)}{\ell(y)} \le \left(\frac{\delta + \|x-y\|}{\delta}\right)^n.$$

Taking the logarithm of both sides,

$$\ln \ell(x) - \ln \ell(y) \le n \ln\left(\frac{\delta + \|x-y\|}{\delta}\right) \le \frac{n\,\|x-y\|}{\delta}.$$

Since the argument is symmetric in x and y, part (ii) of the lemma follows. □

We may now avoid inequality (6.21) by taking a more direct route, which is opened up by replacing Lemma 6.18 by:

Lemma 6.22. *With* $S_0, S_1, \ldots, S_{m-1}$ *as in Lemma 6.18, the sequence*

$$\mu(S_0)^{1/2n},\ \mu(S_1)^{1/2n},\ \ldots,\ \mu(S_{m-1})^{1/2n}$$

is concave. Consequently, the sequence

$$\frac{1}{\mu(S_0)}, \frac{1}{\mu(S_1)}, \ldots, \frac{1}{\mu(S_{m-1})}$$

is convex.

This lemma follows from a functional version of the Brunn-Minkowski Theorem due to Dinghas [21, Satz 1]. We state this theorem in a slightly less general form than it appears in [21].

Theorem 6.23 (Dinghas). *Suppose* A_1 *and* A_2 *are non-empty, bounded, measurable sets in* $\mathbb{R}^n$*, and let* $A_0 = A_1 + A_2$ *be the Minkowski sum of* A_1 *and* A_2*. Suppose* *further that f_1 and f_2 are measurable functions defined on A_1 and A_2, respectively, and form the function g_0 defined by*

$$g_0(x) = \sup\left\{ \left((f_1(x')^{1/r} + f_2(x'')^{1/r} \right)^r : x' \in A_1, x'' \in A_2 \text{ and } x' + x'' = x \right\},$$

for all $x \in A_0$. If f_0 is any measurable function on A_0 satisfying $f_0(x) \geq g_0(x)$ for all $x \in A_0$, then

$$\left[\int_{A_0} f_0(x)\,dx \right]^{1/(r+n)} \geq \left[\int_{A_1} f_1(x)\,dx \right]^{1/(r+n)} + \left[\int_{A_2} f_2(x)\,dx \right]^{1/(r+n)}.$$

Proof of Lemma 6.22. In Theorem 6.23 make the following identifications: $r = n$, $A_1 = S_{i-1}$, $A_2 = S_{i+1}$, $f_1 = f_2 = \ell$ and $f_0(x) = 2^r \ell(x/2)$. By part (i) of Lemma 6.21, we then have $f_0 \geq g_0$, as required; also observe that $2S_i \supseteq S_{i-1} + S_{i+1} = A_0$. The first claim in Lemma 6.22 may then be read off from the concluding inequality of Theorem 6.23. The second claim uses the same reasoning as in the final step of the proof of Lemma 6.18. See also [44, Lemma 2.1]. □

Armed with Lemma 6.22, the upper bound on $a_{i,j}$ derived in the sequence of inequalities ending at (6.22) — with improved constant 1 in place of 2.5 — follows directly from the definition (6.17) of $a_{i,j}$. This establishes Claim 6.16 in the absence of the curvature condition (6.3).

The other place at which the curvature condition is used, namely in establishing (6.29), is trickier to handle. (Note that we used it in going from (6.28) to (6.29).) Our use of curvature is more substantial here, and we need to modify the partitioning of the needle-like body K_1 used in the proof of Claim 6.16 (see Figure 6.5) to recover the proof. If we are prepared to settle for a Poincaré constant λ smaller by a factor n (i.e., $\lambda = c_2\delta^2/D^2n^2$) then it is not too difficult to establish Theorem 6.7 in the absence of (6.3), and we shall see presently how this is done. Getting the correct (up to a constant factor) λ in the absence of (6.3) requires a more complicated analysis, which we only sketch here.

What is it we were trying to achieve with inequality (6.29)? Well, the final contradiction required us to find a set $I \subseteq K$ with the properties that: (i) every point of K_0 is within distance δ of every point of I; and (ii) the ratio $\operatorname{vol}_n I / \operatorname{vol}_n(B(x,\delta) \cap K)$ is bounded below by a universal constant for every $x \in K_0$. Without (6.3) there is currently no guarantee that such a set I exists. However, if we chop K_1 more finely, into slabs of width $\eta = c_3\delta/n$ (instead of $\eta = c_3\delta/\sqrt{n}$), then we are assured to find the required set I. This finer partition increases the number of slabs m by a factor $\sqrt{n}$, and hence reduces the Poincaré constant by a factor n. We borrow the following lemma from Kannan, Lovász and Simonovits [39, Lemma 3.5].

Lemma 6.24. *Suppose $\delta' > 0$, and $x, y \in K$ with $\|x - y\| \le \delta'/\sqrt{n}$. Then*

$$\begin{aligned}&\mathrm{vol}_n(B(x,\delta') \cap B(y,\delta') \cap K)\\ &\quad\ge \frac{1}{1+e}\min\bigl\{\mathrm{vol}_n(B(x,\delta') \cap K),\ \mathrm{vol}_n(B(y,\delta') \cap K)\bigr\}.\end{aligned}$$

Recall that $\mathrm{vol}_n(B(x,\delta') \cap K)$ is proportional to $\ell(x)$. (This is by definition (6.4) of local conductance ℓ.) Now, with η smaller than before, part (ii) of Lemma 6.21 (the Lipschitz inequality for ℓ) ensures that $\mathrm{vol}_n(B(x,\delta') \cap K)$ varies by at most a constant factor as x ranges over K_0. So, choosing δ' a little less than δ, as before, we see that the set $I := B(z_1,\delta') \cap B(z_2,\delta') \cap K$ has the properties we desire: property (i) is by the triangle inequality, and property (ii) is by Lemma 6.24. This establishes Theorem 6.7 without assumption (6.3) but with λ smaller by a factor n.

Exercise 6.25. Flesh out the details of the above proof sketch.

Finally, some inadequate pointers on how to drop assumption (6.3) without losing the factor n in λ. Let's step back and consider what we need to have in order to be able to construct the contradictory set I, using Lemma 6.24. Certainly we need the slabs in the decomposition to have width $O(\delta/\sqrt{n})$; but we also require that the local conductance ℓ varies by at most a constant factor over each slab. As we have seen, these two requirements can be met by using slabs of width $O(\delta/n)$, but then the number of slabs increases, and our estimate of the Poincaré constant worsens.

So it seems that we need to partition K_1 into slabs of unequal thickness, using thinner slabs where ℓ is rapidly varying. We might as well use the coarsest possible partition that will allow us to draw the final contradiction. Starting at the leftmost point of K_1, partition K_1 into slabs $S_0, S_1, \dots, S_{m-1}$ as in Figure 6.5, finishing with slab S_{m-1} at the rightmost point of K_1. Having created $S_0, S_1 \dots, S_{i-1}$, choose the plane defining S_i to be the rightmost plane subject to the conditions:

(i) the distance from the previous plane (i.e., the thickness of slab S_i) is at most $c_3\delta/\sqrt{n}$; and

(ii) the local conductance $\ell(x)$ varies by at most a factor 2 as x ranges over S_i.

Thus the partition of K_1 into slabs S_i is the coarsest possible, subject to conditions (i) and (ii).

Note that conditions (i) and (ii) together allow us to construct, using Lemma 6.24, the set I that leads to the final contradiction. We need of course to fix up the proof of Claim 6.16, which was conducted under the assumption that K_1 is partitioned into slabs of constant width $O(\delta/\sqrt{n})$. Specifically, we need work harder to prove the key inequality (6.24).

Exercise 6.26. Complete the Proof of Theorem 6.7 (the Poincaré inequality) in the absence of the curvature condition (6.3), using the programme outlined above.

The main technical challenge lies in reproving Claim 6.16 in the absence of (6.3), specifically in re-establishing (6.24), taking due account of the amended partition of K_1 into slabs. You will find that the partition of Figure 6.5 (using the amended construction just presented) can be divided into three sections: $S_0, \dots, S_{\ell-1}$, then $S_\ell, \dots, S_{r-1}$ and $S_r \dots, S_{m-1}$, where the slabs in the middle section are all of full width η, and the others are all of strictly smaller width. (Either or both of the outer sections may be empty.) The existence of such a division relies on log-concavity of the local conductance ℓ, which is a consequence of Lemma 6.21(i). The middle section is dealt with exactly as before, since the number of slabs contained within it is $r - \ell \le D/\eta = O(D\sqrt{n}/\delta)$. In the left (right) sections it can be shown that $w_i = \mu(S_i)$ is increasing (decreasing) geometrically; thus the sum (6.17) is determined, up to a constant factor, by its first (last) term. (This step uses log-concavity of ℓ and Brunn-Minkowski.) Thus it doesn't matter so much that the number of terms in the sum (i.e., slabs in the partition) may grow faster than $O(D\sqrt{n}/\delta)$. Note that this is a challenging, verging on speculative, exercise. To keep the technical complexities within bounds, you may want to assume $\delta = O(D/\sqrt{n})$. This is not a restriction in the volume application, where $\delta = \Theta(1/\sqrt{n})$ and $D = \Omega(1)$. However, the assumption is a definite blemish, in that Theorem 6.7 should hold even when δ is of the same order as D.

Remark 6.27. Kannan, Lovász and Simonovits [39] restrict the function f to be an indicator function $f : K \to \{0, 1\}$. The parameter Φ corresponding to λ in the inequality

$$\mathcal{E}_\mu(f, f) \ge \Phi \operatorname{Var}_\mu f, \quad \text{for all (measurable) } f : K \to \{0, 1\}$$

is called the *conductance* of the ball walk. Since the class of functions f is restricted, the conductance Φ is potentially larger than λ. However it is known — a version of Cheeger's inequality — that $\lambda \ge \frac{1}{8}\Phi^2$. (See Sinclair [57] or Aldous and Fill [2] for relationships between various MC parameters, including these two.) The approach to the ball walk in [39] is to show that the conductance Φ is of order $\delta/D\sqrt{n}$, which leads by Cheeger to the required bound on λ. However, the restriction of f to the class of indicator functions unfortunately does not seem to lead to any significant technical simplification in the proof.

6.7 Using samples to estimate volume

In order to estimate the volume of a convex body using our sampling procedure, we follow the basic "product of ratios" approach used in earlier examples. Briefly, the procedure is as follows.

Given our convex body K, we define a series of concentric balls $B_0 \subset B_1 \subset \cdots \subset B_k$ such that $B_0 \subseteq K$ and $K \subseteq B_k$. (Refer to Figure 6.10.) Additionally, we

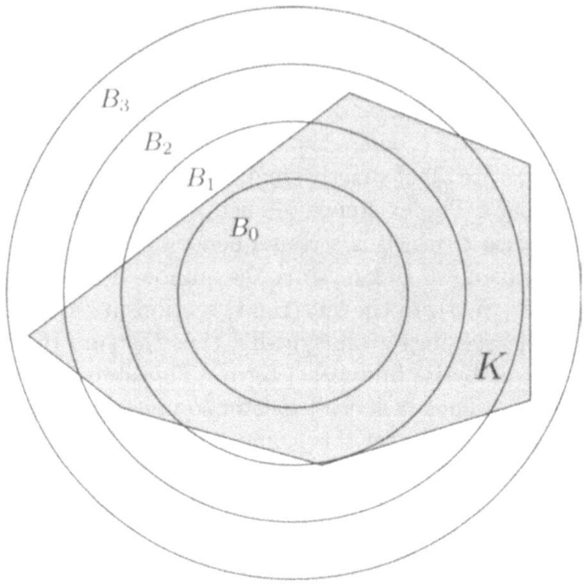

Figure 6.10: convex body K and the concentric balls

require that the volume of these balls does not grow too quickly, say $\mathrm{vol}_n\, B_{i+1} \leq 2\,\mathrm{vol}_n\, B_i$. We can estimate the ratios

$$\varrho_i = \frac{\mathrm{vol}_n(B_i \cap K)}{\mathrm{vol}_n(B_{i+1} \cap K)}$$

by repeatedly sampling points from $B_{i+1} \cap K$ and determining the fraction of these points which lie also in $B_i \cap K$. Let Z_i be an estimate for ϱ_i obtained by taking the sample mean. We then get the desired estimate of $\mathrm{vol}_n\, K$ from

$$\mathrm{vol}_n\, K \approx \mathrm{vol}_n\, B_0 \cdot \prod_{i=0}^{k-1} \frac{1}{Z_i}.$$

Of course, we may calculate $\mathrm{vol}_n\, B_0$ from an explicit formula.

We have glossed over important issues here, not least the obvious fact that k must not be too large if we are to control the variance of our product estimator for $\mathrm{vol}_n\, K$. If K is "well rounded" then, indeed, k need not be very large. But if K is very elongated it will be necessary to apply a linear transformation to K to render it well rounded. For details of this step, and many further refinements, refer to [39].

6.8 Appendix: a proof of Corollary 6.8

We work with the lazy version of the ball walk, which stays put with probability $\frac{1}{2}$. For the first leg, we follow closely the proof of Theorem 5.6, but replacing sums by integrals. Because of the close similarity of the arguments we record only the main steps here:

$$[P_{\text{zz}}f](x) = \frac{1}{2}\int_K P(x,dy)\,\big(f(x)+f(y)\big),$$
$$\operatorname{Var}_\mu(P_{\text{zz}}f) \le \frac{1}{4}\int_K \mu(dx)\int_K P(x,dy)\big(f(x)+f(y)\big)^2,$$

and

$$\operatorname{Var}_\mu f = \frac{1}{2}\int_K \mu(dx)\int_K P(x,dy)\big(f(x)^2+f(y)^2\big).$$

It follows that

$$\begin{aligned}\operatorname{Var}_\mu f - \operatorname{Var}_\mu(P_{\text{zz}}f) &\ge \frac{1}{4}\int_K \mu(dx)\int_K P(x,dy)\big(f(x)-f(y)\big)^2\\ &= \frac{1}{2}\mathcal{E}_\mu(f,f)\\ &\ge \frac{1}{2}\lambda\operatorname{Var}_\mu f,\end{aligned}$$

and hence

$$\operatorname{Var}_\mu(P_{\text{zz}}f) \le \Big(1-\frac{\lambda}{2}\Big)\operatorname{Var}_\mu f.$$

Iterating the above, we obtain

$$\operatorname{Var}_\mu(P_{\text{zz}}^t f) \le \Big(1-\frac{\lambda}{2}\Big)^t\operatorname{Var}_\mu f \le \exp(-\tfrac{1}{2}\lambda t). \tag{6.34}$$

Now suppose A is measurable subset of K, and let $f : K \to \mathbb{R}$ be the function that is 1 on A and 0 outside A. Assume that we start our walk from a point X_0 selected uniformly at random from the ball $B = B(x,\delta) \subseteq K$. (This is, of course, equivalent to starting the walk at point x at time -1.) For $\varepsilon > 0$ we want to find a time t such that the variation distance of the t-step distribution from stationarity is at most ε; equivalently, we require

$$|\Pr(X_t \in A) - \mu(A)| = \left|\frac{1}{\operatorname{vol}_n B}\int_B \big\{[P_{\text{zz}}^t f](y) - \mu(A)\big\}\,dy\right| \le \varepsilon, \tag{6.35}$$

uniformly over the choice of A. (In this context, recall the definition of total variation distance (3.2), and the fact that $[P_{\text{zz}}^t f](y)$ may be interpreted as $\Pr(X_t \in A \mid X_0 = y)$.)

Noting $\mathrm{E}_\mu(P^t_{\mathrm{zz}}f) = \mu(A)$, we find

$$\begin{aligned}
\operatorname{Var}_\mu(P^t_{\mathrm{zz}}f) &\geq \int_B \big\{[P^t_{\mathrm{zz}}f](y) - \mu(A)\big\}^2 \mu(dy) \\
&\geq \frac{0.4}{\mathrm{vol}_n K} \int_B \big\{[P^t_{\mathrm{zz}}f](y) - \mu(A)\big\}^2 dy \qquad (6.36)\\
&\geq \frac{0.4\,\mathrm{vol}_n B}{\mathrm{vol}_n K} \left[\frac{1}{\mathrm{vol}_n B} \int_B \big\{[P^t_{\mathrm{zz}}f](y) - \mu(A)\big\}\, dy\right]^2, \qquad (6.37)
\end{aligned}$$

where inequality (6.36) follows from the definition (6.5) of μ and Lemma 6.4; and (6.37) from the fact that the expectation of the square of a r.v. is at least as large as the square of its expectation. Thus, to achieve the desired bound (6.35) on variation distance, we require

$$\operatorname{Var}_\mu(P^t_{\mathrm{zz}}f) \leq \frac{0.4\,\varepsilon^2\,\mathrm{vol}_n B}{\mathrm{vol}_n K}.$$

Now, the volume of K is maximised, for specified diameter D, when K is a ball of radius $D/2$. Thus it is enough that we achieve

$$\operatorname{Var}_\mu(P^t_{\mathrm{zz}}f) \leq 0.4\,\varepsilon^2 \left(\frac{2\delta}{D}\right)^n.$$

According to (6.34), this inequality will hold, provided

$$t \geq \left\lceil \frac{2}{\lambda}\left(\ln\left\{\frac{5}{2\varepsilon^2}\right\} + n\ln\left\{\frac{D}{2\delta}\right\}\right)\right\rceil.$$

This is the mixing time claimed in Corollary 6.8, with $i(\mu_0)$ specialised to an initial distribution that is uniform and supported on a ball of radius δ.

Chapter 7

Inapproximability

Not all counting problems are efficiently approximable. We open with a simple example.

Fact 7.1. *Unless* RP $=$ NP *there can be no FPRAS for the number of Hamilton cycles in a graph G.*

Informally: assuming, as seems likely, that there exist predicates in NP that admit no polynomial-time randomised algorithm, then no FPRAS for Hamilton cycles can exist. Still informally: the reason is that an FPRAS for Hamilton cycles would, in particular, need to distinguish the zero from non-zero case with reasonable probability.

To apply a rigorous interpretation to Fact 7.1, we need to divert briefly into randomised complexity classes, in particular RP and BPP. A predicate $\varphi : \Sigma^* \to \{0,1\}$ is in the class RP if there is a polynomial-time witness-checking predicate[1] $\chi : \Sigma^* \times \Sigma^* \to \{0,1\}$ and a polynomial p such that:

(i) if $\neg\varphi(x)$ then $\neg\chi(x,w)$ for all putative witnesses $w \in \Sigma^{p(|x|)}$;

(ii) if $\varphi(x)$ then $\Pr[\chi(x,w)] \geq \frac{1}{2}$, where w is assumed to be chosen u.a.r. from the set $\Sigma^{p(|x|)}$.

The predicate φ is in the class BPP if there exist χ and p, as above, satisfying:

(i′) if $\neg\varphi(x)$ then $\Pr[\chi(x,w)] \leq \frac{1}{4}$;

(ii′) if $\varphi(x)$ then $\Pr[\chi(x,w)] \geq \frac{3}{4}$,

where, again, w is assumed to be chosen u.a.r. from the set $\Sigma^{p(|x|)}$. Thus RP (respectively, BPP) is the set of predicates that can be decided in randomised polynomial time with one-sided (respectively, two-sided) error.

[1]Refer to Chapter 2 for the general setting.

Remarks 7.2. (a) There is no significance in the exact thresholds $\frac{1}{2}$, $\frac{1}{4}$ and $\frac{3}{4}$ appearing in the above definitions. By designing appropriate simulations, one can show that $\frac{1}{2}$ can be replaced by any constant strictly between 0 and 1, and $\frac{1}{4}$ and $\frac{3}{4}$ by any constants c_1, c_2 with $0 < c_1 < c_2 < 1$.

(b) It is immediate from the definition of RP that RP $\subseteq$ NP. No similar inclusion is known for BPP.

Now, comparing the definition of BPP with that of FPRAS, we see that the existence of an FPRAS for the number of Hamilton cycles in a graph G would immediately imply that the decision problem — is G Hamiltonian? — is in BPP. Since the decision problem is NP-complete, it would follow that NP $\subseteq$ BPP. The apparently stronger conclusion RP = NP follows from the complexity-theoretic fact:

Fact 7.3. *If* NP $\subseteq$ BPP *then* NP $\subseteq$ RP *(and hence* RP = NP*).*

See, e.g., Papadimitriou's textbook [54, Problem 11.5.18].

Remark 7.4. The converse to Fact 7.1 is also true: if RP = NP then there is an FPRAS for the number of Hamilton cycles in a graph. Whereas Fact 7.1 is trivial, its converse is not, relying as it does on the bisection method of Valiant and Vazirani [63]. See Chapter 10 of Goldreich's lecture notes [32].

Of course, Hamiltonicity is not a distinguished NP-complete problem. More generally we have:

Fact 7.5. *(Informal statement.) If the decision version of a counting problem is* NP*-complete, then the counting problem itself does not admit an FPRAS unless* RP = NP.

Exercise 7.6. Provide a formal statement of Fact 7.5 using the notion of witness-checking predicates.

Fact 7.5 instantly yields a large number of counting problems that, for a rather trivial reason, do not admit an FPRAS (under a reasonable complexity-theoretic assumption). We now turn to an example that does not admit an FPRAS for some non-trivial (though only slightly non-trivial) reason.

Let us consider the independent sets counting problem:

Name. #IS.

Instance. A graph G.

Output. The number of independent sets[2] of all sizes in G.

[2] An independent set in graph G is a subset $U \subseteq V(G)$ of the vertex set of G such that no edge of G has both endpoints in U.

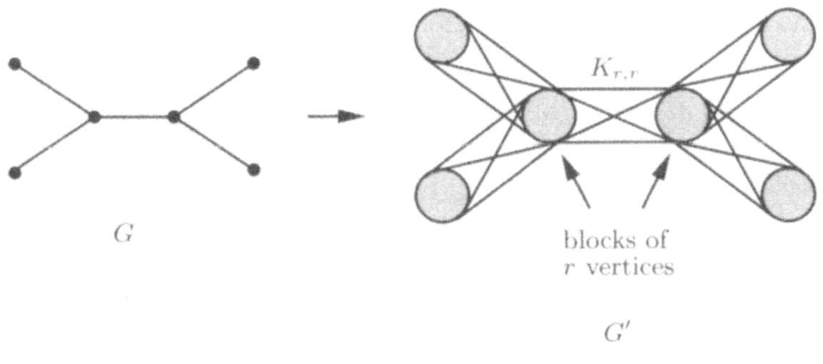

Figure 7.1: The construction.

The decision version of #IS is trivial, since every graph has the empty set of vertices as an independent set. Nevertheless, we shall see that #IS is hard to approximate under some reasonable complexity-theoretic assumption. We shall make use of the optimisation version of #IS:

Name. MAXIS.

Instance. A graph G.

Output. The size of a maximum independent set in G.

MAXIS is a classical NP-complete[3] problem: see, e.g., Garey and Johnson [30, GT20].

Proposition 7.7. *There is no FPRAS for* #IS *unless* RP = NP.

Proof. We use a reduction from MAXIS. Let $G = (V, E)$ be an instance of MAXIS. We want to construct a graph $G' = (V', E')$, being an instance of #IS, in such a way that *typical* independent sets in G' reveal *maximum* independent sets in G.

The construction replaces vertices by blocks of r vertices and edges by complete bipartite graphs between blocks; formally,

$$V' = V \times \{0, \dots, r-1\},$$

and

$$E' = \big\{\{(u,i),(v,j)\} : \{u,v\} \in E \text{ and } i,j \in \{0,\dots,r-1\}\big\}.$$

(See Figure 7.1.)

[3]To make formal sense of this claim, one would need to make MAXIS into a decision problem. This could be done, in the usual way, by specifying a bound $k \in \mathbb{N}$ as part of the problem instance and asking whether G has an independent set of size at least k.

Each independent set I' in G' projects to an independent set

$$I = \{v \in V : \text{there exists } i \in \{0, \ldots, r-1\} \text{ such that } (v, i) \in I'\}$$

in G. (Since each edge of G corresponds to a complete bipartite subgraph in G', the set I is indeed independent in G.) Suppose $|I| = k$; then there are $(2^r - 1)^k$ independent sets I' in G' that project to the specific independent set I in G. We consider the two complementary situations:

(a) An independent set of size k exists in G. Then there are at least $(2^r - 1)^k$ independent sets in G'.

(b) The maximum independent set in G has size less than k. Then there are at most $2^n(2^r - 1)^{k-1}$ independent sets in G', where $n = |V|$.

Setting $r = n + 2$, we have

$$(2^r - 1)^k = (2^{n+2} - 1)(2^r - 1)^{k-1} \geq 2 \times 2^n(2^r - 1)^{k-1};$$

in other words, the minimum possible number of independent sets in case (a) exceeds the maximum possible number in case (b) by a factor 2. An FPRAS for #IS would be able to distinguish cases (a) and (b) with high probability, providing us with a polynomial-time randomised algorithm (with two-sided error) for MAXIS. As we have seen, this would imply RP = NP. □

Remark 7.8. Note that the reduction proves something much stronger than the non-existence of an FPRAS for #IS. It shows (under the assumption RP $\neq$ NP) that there in no polynomial time randomised algorithm that approximates the number of independent sets even to within any fixed exponential factor. To see this, simply set $r = cn$ with $c > 1$. The statement can be strengthened even further: see Dyer, Frieze and Jerrum [22].

7.1 Independent sets in a low degree graph

Proposition 7.7 is evidence that the number of independent sets in a graph is hard to approximate in general, so we need to restrict the class of problem instances to make progress. One interesting way to do this is to place a bound Δ on the maximum degree of the instance G. Then we can investigate how the computational difficulty of of #IS varies as Δ does. On the positive side we have the following result.

Theorem 7.9 (Luby and Vigoda). *There is an FPRAS for #IS when $\Delta = 4$.*

Proof (sketch). As usual, it is enough to be able to sample independent sets almost uniformly at random in polynomial time.

Independent sets are sampled using an MC based on edge updates. View an independent set I in graph $G = (V, E)$ as a function $I : V \to \{0, 1\}$, where $I(v) = 1$

has the interpretation that v is in the independent set. The state space of the MC is the set of all independent sets in G. Transition probabilities are specified by the following trial, where $X_0 : V \to \{0,1\}$ is the initial independent set.

1. Choose an edge $\{u,w\} \in E$, u.a.r.

2. Begin to construct a new independent set I as follows: with equal probability ($\frac{1}{3}$ in each case) set (a) $I(u) := 0$ and $I(w) := 0$; (b) $I(u) := 0$ and $I(w) := 1$; or (c) $I(u) := 1$ and $I(w) := 0$. (Note that these three cases correspond to the three possible restrictions of an independent set in G to the edge $\{u,w\}$.)

3. For all $v \in V \setminus \{u,w\}$ set $I(v) := X_0(v)$.

4. If I is an independent set then $X_1 := I$, otherwise $X_1 := X_0$.

Informally, we are using edge-updates with Metropolis acceptance probabilities.

This MC can be shown to be rapidly mixing using the path-coupling method. Two independent sets are considered to be adjacent if they differ at exactly one vertex. If adjacent independent sets are considered to be at distance 1, the derived path-metric is just Hamming distance. Suppose X_0 and Y_0 are adjacent; on the basis of a case analysis of moderate complexity it is possible to conclude that the expected Hamming distance between X_1 and Y_1 is at most 1. (For a regular graph with no small cycles there are four "good edges" $\{u,w\}$ whose selection may cause the distance to decrease, and twelve "bad edges" which may cause the distance to increase. In the worst case, these two effects are exactly in balance.) It follows that the mixing time of the MC scales quadratically with n. □

Exercise 7.10. Complete the proof of Theorem 7.9. To keep technical complexity to a minimum, assume the graph G is triangle-free, i.e., contains no cycles of length 3. In case you need to refer to it, a complete analysis (in a more general setting where vertices in the independent set are given weight or "fugacity" λ) is given by Luby and Vigoda [47]. Theorem 7.9 corresponds to the case $\lambda = 1$ of their result. Dyer and Greenhill [25] also obtain a generalisation of Theorem 7.9, using a slightly different MC. Their proof has the advantage of dispensing with triangle-freeness.

According to Theorem 7.9, approximately counting independent sets in a graph G is tractable provided the maximum degree Δ is small enough. We know that $\Delta = 4$ is small enough, so what about $\Delta = 5, 6, \ldots$? The reduction described in Proposition 7.7 constructs graphs of arbitrarily large degree, so it apparently leaves open the possibility that there is an FPRAS for #IS for any fixed degree bound Δ. However, if we look afresh at the construction of Theorem 7.9 in the light of inapproximability results for the optimisation problem MaxIS, we discover that there is a definite upper bound on Δ. This idea is due to Luby and Vigoda [47].

Proposition 7.11. *There is no FPRAS for* #IS *when* $\Delta = 1188$, *unless* RP = NP.

Proof. We know that MAXIS is NP-hard when restricted to graphs of maximum degree 4. A result of Berman and Karpinski [6, Thm 1(iv)] tells us more: for any $\varepsilon > 0$, it is NP-hard to determine the size of a maximum independent set in a graph G to within ratio of $\frac{73}{74} + \varepsilon$, even when G is restricted to have maximum degree 4. (By "determining the size… within ratio ϱ" we mean computing a number $\hat{k}$ such that $\varrho k \leq \hat{k} \leq k$, where k is the size of a maximum independent set in G.) In other words, the problem MAXIS is polynomial-time (Turing) reducible to the approximate version of MAXIS, in which we ask for a result within ratio $\frac{73}{74} + \varepsilon$. This result, like many other inapproximability results for optimisation problems, rests on the powerful theory of probabilistically checkable proofs (PCP).

So let G be a graph of maximum degree 4. Using our construction from the proof of Theorem 7.7 with $r = 297$, we obtain a graph G' of maximum degree 1188. We shall see that even a rough approximation to the *number* of independent sets in G' will provide a close (within ratio $\frac{73}{74} + \varepsilon$) approximation to the *size* of the largest independent set in G. Thus the existence of an FPRAS for #IS in graphs of maximum degree 1188 would imply the existence of a polynomial-time randomised algorithm (with two-sided error) for MAXIS. As before, this would in turn imply RP = NP.

We define J' to be the collection of all independent sets in G'. Let k be the size of a maximum independent set in G. We have

$$(2^r - 1)^k \leq |J'| \leq 2^n (2^r - 1)^k,$$

or, taking the natural logarithm,

$$k \ln(2^r - 1) \leq \ln |J'| \leq n \ln 2 + k \ln(2^r - 1).$$

Consider the following estimate for k:

$$\hat{k} = \frac{\ln |J'| - n \ln 2}{\ln(2^r - 1)};$$

it is clear that

$$k - \frac{n \ln 2}{\ln(2^r - 1)} \leq \hat{k} \leq k.$$

Recall that Brooks's theorem [7, 9] asserts that any graph of maximum degree $\Delta \geq 3$ that does not contain $K_{\Delta+1}$ as a connected component is Δ-colourable. Assuming, as we may, that G is connected, it follows that G is 4-colourable. Since any (and hence in particular the largest) of the four colour classes is an independent set, $k \geq n/4$. Thus

$$k \left(1 - \frac{4 \ln 2}{\ln(2^r - 1)}\right) \leq \hat{k} \leq k.$$

Note that, when $r = 297$,

$$\frac{4 \ln 2}{\ln(2^r - 1)} < \frac{1}{74}.$$

If we had an FPRAS for #IS restricted to graphs of maximum degree 1188 then we would be able to approximate $|J'|$ (with high probability) within arbitrarily small constant relative error, and $\ln|J'|$ (and hence $\hat{k}$) within arbitrarily small constant additive error. But this in turn would provide an approximation to the size of the largest independent set in G (with high probability) within ratio $\frac{73}{74} + \varepsilon$. □

One might suspect that the degree bound $\Delta = 1188$ in Proposition 7.11 is quite a bit larger than necessary, and this is indeed the case. Indeed, simply by tightening the analysis of the construction used in the proof of Proposition 7.11, one can reduce the degree Δ in its statement by 10–20%.

Exercise 7.12. Using the same reduction, but improved estimates, show that Proposition 7.11 holds for some Δ less than 1100. (I think $\Delta = 964$ is achievable.)

Using a technically more involved reduction, Dyer, Frieze and Jerrum have shown that $\Delta = 1188$ may be replaced by $\Delta = 25$. That still leaves a large gap between what is known to be tractable ($\Delta = 4$) and intractable ($\Delta = 25$); no doubt the upper bound could be reduced slightly at the expense of additional technical complexity, but a substantial gap would still remain.

To explore further the boundary between tractable and intractable requires us, at present, to accept more circumstantial evidence. Consider any MC on independent sets of a graph on n vertices. Let $b(n) \leq n$ be any function of n and suppose the Hamming distance between successive states X_t and X_{t-1} of the MC is uniformly bounded by $b(n)$. We will say that the MC is $b(n)$-*cautious.* (Recall that we are viewing independent sets as functions $V \to \{0,1\}$.) Thus a $b(n)$-cautious MC is not permitted to change the status of more than $b(n)$ vertices in G at any step. Ideally, for ease of implementation, we would wish to have $b(n)$ a constant (as in the proposals of Luby and Vigoda [47], and Dyer and Greenhill [25]). However, we are able show that no $b(n)$-cautious chain on independent sets can mix rapidly unless $b(n) = \Omega(n)$, even when $\Delta = 6$. Thus any chain that *does* mix rapidly on graphs of maximum degree 6 must change the status of a sizeable proportion of the vertices at each step.

Theorem 7.13 (Dyer, Frieze and Jerrum). *There exists an infinite family of regular bipartite graphs of degree* 6, *together with constants* $\delta, \gamma > 0$, *such that the following is true: any* δn-*cautious MC on independent sets of these graphs has exponential mixing time, in the sense that* $\tau(\frac{1}{4}) = \Omega(\exp(\gamma n))$.

Dyer, Frieze and Jerrum's proof of Theorem 7.13 provides an explicit value for δ, namely $\delta = 0.35$. We present a simplified version of the proof here that does not attempt to estimate δ. The idea underlying the proof is very simple: if the state space of an MC has a tight "constriction" then its mixing time will be long. This intuition may be formalised as follows.

Claim 7.14. *Consider an MC with state space Ω, transition matrix P, and stationary distribution π. Let $A \subset \Omega$ be a set of states such that $\pi(A) \leq \frac{1}{2}$, and $M \subset \Omega$ be a set of states that forms a "barrier" in the sense that $P(i,j) = 0$ whenever $i \in A \setminus M$ and $j \in \overline{A} \setminus M$. Then the mixing time τ of the MC satisfies $\tau(\frac{1}{4}) \geq \pi(A)/4\pi(M)$.*

We defer the proof of the claim to the end of the chapter.

Proof of Theorem 7.13. Our counterexample to rapid mixing (or, more precisely, family of counterexamples indexed by n) is a random regular bipartite graph G of degree $\Delta = 6$, with n vertices on the left and n on the right. Denote the left and right vertex sets by V_1 and V_2 respectively. The random graph model is simple. A *pairing* is one of the $n!$ possible bijections between left and right vertices viewed as a regular bipartite graph of degree 1. Select Δ pairings, independently and u.a.r., and form the union: the result is a bipartite graph G of maximum degree Δ. Since the pairings may not be disjoint, the graph G may not be regular; we return to this point later.

Let $J(\alpha,\beta)$ be the collection of all independent sets in G having αn vertices on the left and βn on the right. For a given set of αn vertices $U_1 \subseteq V_1$ and βn vertices $U_2 \subseteq V_2$, what is the probability that a random pairing will avoid joining some element in U_1 to some element in U_2? Well, the "image" of U_1 under the pairing is a random αn-subset of V_2, so the answer is the same as the probability that a random αn-subset of V_2 is disjoint from U_2; but the latter probability is just

$$\binom{(1-\beta)n}{\alpha n} \Big/ \binom{n}{\alpha n}.$$

Thus the expected size of $J(\alpha,\beta)$ for a random G chosen according to our model is just

$$\mathrm{E}\,|J(\alpha,\beta)| = \binom{n}{\alpha n}\binom{n}{\beta n}\left[\binom{(1-\beta)n}{\alpha n} \Big/ \binom{n}{\alpha n}\right]^{\Delta}.$$

(By linearity of expectation, the required quantity is simply the number of possible candidates (U_1, U_2), times the probability that all Δ pairings avoid connecting U_1 and U_2.) By Stirling's approximation we have

$$\mathrm{E}\,|J(\alpha,\beta)| = \exp\big(\varphi(\alpha,\beta)\, n(1+o(1))\big)$$

where

$$\begin{aligned}\varphi(\alpha,\beta) &= -\alpha\ln\alpha - \beta\ln\beta - \Delta(1-\alpha-\beta)\ln(1-\alpha-\beta) \\ &\quad + (\Delta-1)\big((1-\alpha)\ln(1-\alpha) + (1-\beta)\ln(1-\beta)\big). \qquad (7.1)\end{aligned}$$

We treat φ as a function of real arguments α and β, even though a combinatorial interpretation is possible only when αn and βn are integers. Then φ is defined on the triangle

$$\mathcal{T} = \{(\alpha,\beta) : \alpha,\beta \geq 0 \text{ and } \alpha+\beta \leq 1\},$$

and is clearly symmetrical in α, β. (The function φ is defined by equation (7.1) on the interior of $\mathcal{T}$, and can be extended to the boundary by taking limits.)

Now set $\Delta = 6$. By calculus, $\varphi(\alpha, \alpha)$ has a unique maximum in the range $[0, \frac{1}{2})$; numerically $\varphi(\alpha, \alpha)$ is uniformly less than 0.704 in this range. Consider the region $\mathcal{D} = \{(\alpha, \beta) \in \mathcal{T} : |\alpha - \beta| \leq \delta\}$, where δ is a small positive constant. (This is the δ in the statement of the theorem.) For sufficiently small $\delta > 0$,

$$\varphi(\alpha, \beta) \leq 0.705, \quad \text{for all } (\alpha, \beta) \in \mathcal{D}.$$

For, if not, there would be an infinite sequence (α_i, β_i) of points in $\mathcal{T}$, all satisfying $\varphi(\alpha, \beta) > 0.705$, which approach the diagonal $\alpha = \beta$ arbitrarily closely. By compactness, there would be a subsequence of (α_i, β_i) converging to some point on the diagonal, contradicting continuity of φ. So, by Markov's inequality, with very high probability,[4]

$$\left| \bigcup_{(\alpha,\beta)\in\mathcal{D}} J(\alpha, \beta) \right| \leq e^{0.706n}, \tag{7.2}$$

where the union is over α, β which are multiples of $1/n$.

Denote by $\mathcal{L}$ and $\mathcal{R}$ the two connected regions of $\mathcal{T} \setminus \mathcal{D}$. We need a lower bound on the number of independent sets in these regions which exceeds the upper bound (7.2). With this in mind, define

$$\theta(\alpha) = -\alpha \ln \alpha - (1 - \alpha) \ln(1 - \alpha) + (\ln 2)(1 - \Delta\alpha).$$

for $\alpha < \Delta^{-1}$. Then, for *any* graph G in the space of random graphs, the total number of independent sets I with $|I \cap V_1| = \alpha n$ is (crudely) at least

$$|J(\alpha, *)| \geq \binom{n}{\alpha n} 2^{(1-\Delta\alpha)n} = \exp\big(\theta(\alpha)\, n(1 - o(1))\big).$$

(Choose αn vertices from V_1; then choose any subset of vertices from the at least $(1 - \Delta\alpha)n$ unblocked vertices in V_2.) Set $\Delta = 6$ as before and $\alpha^* = 0.015$. Then, by numerical computation, $\theta(\alpha^*)$ is greater than 0.708. In other words,

$$\left| \bigcup_{(\alpha,\beta)\in\mathcal{L}} J(\alpha, \beta) \right| \geq e^{0.708n}, \tag{7.3}$$

for all sufficiently large n, with a similar bound for $\mathcal{R}$. Comparing (7.2) and (7.3), we see that, with very high probability, the number of approximately balanced independent sets is smaller, by an exponential factor, than the number with a sizeable imbalance in either direction. Specifically, the former is smaller than the latter by a factor $e^{\gamma n}$, where $\gamma = 0.002$.

[4] "With very high probability" may be taken to mean "with probability differing from 1 by an amount decaying exponentially fast with n."

The $(n+n)$-vertex graph whose existence is guaranteed by Theorem 7.13 (ignoring for a moment the regularity requirement) is any graph from the space of random graphs under consideration that exhibits the exponential gap just described. (A randomly chosen graph will do with high probability.) The remainder of our argument concerns such a graph.

Now consider a δn-cautious MC. Let $A = \bigcup_{\alpha \geq \beta} J(\alpha, \beta)$ denote the set of leftward leaning independent sets, and assume, without loss of generality, that A is no larger than its complement $\overline{A} = \Omega \setminus A$. Denote by M the set of approximately balanced independent sets $M = \bigcup_{(\alpha,\beta)\in\mathcal{D}} J(\alpha, \beta)$.

Since the MC is δn-cautious, it cannot make a transition from A to $\overline{A}$ directly, but only by using intermediate states in M. Now, we know from inequalities (7.2) and (7.3) that

$$|A| \geq e^{\gamma n} |M|. \tag{7.4}$$

If we are prepared to weaken the theorem slightly by dropping the condition that the graphs be regular, we can immediately complete the proof by appealing to Claim 7.14.

We may address the regularity issue by reference to a standard result about the union-of-pairings model for random bipartite graphs. Provided Δ is taken as constant, Bender [5] has shown that Δ-regular graphs occur in our random graph model with probability bounded away from 0. Since we prove that random graphs of maximum degree 6, with very high probability, have the property we seek, it follows that random Δ-regular graphs (in the induced probability space), with very high probability, have the property too. □

It only remains to present the missing proof.

Proof of Claim 7.14. Denote by π_t the t-step distribution of the MC. First note that

$$\begin{aligned}\|\pi_{t+1} - \pi_t\|_{\mathrm{TV}} = \|\pi_t P - \pi_{t-1} P\|_{\mathrm{TV}} &= \frac{1}{2} \max_{\|z\|_\infty \leq 1} (\pi_t - \pi_{t-1}) P z \\ &\leq \frac{1}{2} \max_{\|w\|_\infty \leq 1} (\pi_t - \pi_{t-1}) w \\ &= \|\pi_t - \pi_{t-1}\|_{\mathrm{TV}},\end{aligned}$$

since $\|Pz\|_\infty \leq \|z\|_\infty$. Hence, by induction, $\|\pi_{t+1} - \pi_t\|_{\mathrm{TV}} \leq \|\pi_1 - \pi_0\|_{\mathrm{TV}}$ and, further, using the triangle inequality, $\|\pi_t - \pi_0\|_{\mathrm{TV}} \leq t\,\|\pi_1 - \pi_0\|_{\mathrm{TV}}$. Now, for $\emptyset \subset S \subset \Omega$, define

$$\Phi(S) = \frac{1}{\pi(S)} \sum_{i\in S} \sum_{j \in \overline{S}} \pi(i) P(i,j).$$

The quantity $\Phi = \min\{\Phi(S) : S \subset \Omega \text{ and } 0 < \pi(S) \leq \frac{1}{2}\}$ is sometimes called the "conductance" of the MC. (Conductance is normally considered in the context of time-reversible Markov chains. However, both the definition and the line of argument employed here apply to non-time-reversible chains.) Now

$$\begin{aligned}\sum_{i\in A}\sum_{j\in\overline{A}}\pi(i)P(i,j) &\le \sum_{i\in A}\sum_{j\in\overline{A}\cap M}\pi(i)P(i,j) + \sum_{i\in A\cap M}\sum_{j\in\overline{A}}\pi(i)P(i,j)\\ &\le \pi(\overline{A}\cap M)+\pi(A\cap M)\\ &= \pi(M).\end{aligned}$$

In short, $\Phi(A)\,\pi(A)\le\pi(M)$. So setting

$$\pi_0(i)=\begin{cases}\pi(i)/\pi(A), & \text{if } i\in A;\\ 0, & \text{otherwise,}\end{cases}$$

we have

$$\begin{aligned}\|\pi_1-\pi_0\|_{\mathrm{TV}} &= \frac{1}{2}\sum_{j\in\Omega}\left|\sum_{i\in\Omega}\pi_0(i)P(i,j)-\pi_0(j)\right| &(7.5)\\ &= \sum_{j\in\overline{A}}\sum_{i\in A}\pi_0(i)P(i,j) &(7.6)\\ &= \Phi(A).\end{aligned}$$

(To see equality (7.6), observe that the terms in (7.5) with $j\in A$ make a contribution to the sum that is equal to that made by the terms with $j\in\overline{A}$. Now simply restrict the sum to terms with $j\in\overline{A}$.) But $\|\pi_0-\pi\|_{\mathrm{TV}}\ge\frac{1}{2}$, since $\pi(A)\le\frac{1}{2}$, and hence

$$\|\pi_t-\pi\|_{\mathrm{TV}}\ge\|\pi_0-\pi\|_{\mathrm{TV}}-\|\pi_t-\pi_0\|_{\mathrm{TV}}\ge\frac{1}{2}-t\,\Phi(A).$$

Thus we cannot achieve $\|\pi_t-\pi\|_{\mathrm{TV}}\le\frac{1}{4}$ until

$$t\ge\frac{1}{4\Phi(A)}\ge\frac{\pi(A)}{4\pi(M)}.$$

By an averaging argument there must exist some initial state $x_0\in A$ for which $\tau_{x_0}\left(\frac{1}{4}\right)\ge\pi(A)/4\pi(M)$. □

Bibliography

[1] David Aldous. Random walks on finite groups and rapidly mixing Markov chains. In *Seminar on probability, XVII*, pages 243–297. Springer, Berlin, 1983.

[2] David Aldous and James Fill. Reversible Markov chains and random walks on graphs.
`http://www.stat.berkeley.edu/~aldous/book.html`.

[3] V. S. Anil Kumar and H. Ramesh. Coupling vs. conductance for the Jerrum-Sinclair chain. *Random Structures Algorithms*, 18(1):1–17, 2001.

[4] Catherine Bandle. *Isoperimetric Inequalities and Applications.* Pitman (Advanced Publishing Program), Boston, Mass., 1980.

[5] Edward A. Bender. The asymptotic number of non-negative integer matrices with given row and column sums. *Discrete Math.*, 10:217–223, 1974.

[6] Piotr Berman and Marek Karpinski. On some tighter inapproximability results (extended abstract). In *Automata, languages and programming (Prague, 1999)*, pages 200–209. Springer, Berlin, 1999.

[7] Béla Bollobás. *Modern Graph Theory.* Springer-Verlag, New York, 1998.

[8] Andrei Z. Broder. How hard is it to marry at random? (on the approximation of the permanent). In *Proceedings of the 18th Annual ACM Symposium on Theory of Computing (STOC)*, pages 50–58. ACM Press, 1986. Erratum in *Proceedings of the 20th Annual ACM Symposium on Theory of Computing*, 1988, p. 551.

[9] R. L. Brooks. On colouring the nodes of a network. *Proc. Cambridge Philos. Soc.*, 37:194–197, 1941.

[10] Russ Bubley and Martin Dyer. Graph orientations with no sink and an approximation for a hard case of #SAT. In *Proceedings of the Eighth Annual ACM-SIAM Symposium on Discrete Algorithms (New Orleans, LA, 1997)*, pages 248–257, New York, 1997. ACM.

[11] Russ Bubley and Martin Dyer. Path coupling: a technique for proving rapid mixing in Markov chains. In *Proceedings of the 38th Symposium on Foundations of Computer Science (FOCS)*, pages 223–231. IEEE Computer Society Press, 1997.

[12] Russ Bubley and Martin Dyer. Path coupling, Dobrushin uniqueness, and approximate counting. Technical Report 97.04, School of Computer Studies, University of Leeds, January 1997.

[13] Russ Bubley and Martin Dyer. Faster random generation of linear extensions. *Discrete Math.*, 201(1-3):81–88, 1999.

[14] Russ Bubley, Martin Dyer, and Catherine Greenhill. Beating the 2Δ bound for approximately counting colourings: a computer-assisted proof of rapid mixing. In *Proceedings of the Ninth Annual ACM-SIAM Symposium on Discrete Algorithms (San Francisco, CA, 1998)*, pages 355–363, New York, 1998. ACM Press.

[15] Russ Bubley, Martin Dyer, and Mark Jerrum. An elementary analysis of a procedure for sampling points in a convex body. *Random Structures Algorithms*, 12(3):213–235, 1998.

[16] Charles J. Colbourn, J. Scott Provan, and Dirk Vertigan. A new approach to solving three combinatorial enumeration problems on planar graphs. *Discrete Appl. Math.*, 60(1-3):119–129, 1995.

[17] Colin Cooper and Alan M. Frieze. Mixing properties of the Swendsen-Wang process on classes of graphs. *Random Structures Algorithms*, 15(3-4):242–261, 1999.

[18] Persi Diaconis. *Group representations in probability and statistics.* Institute of Mathematical Statistics, Hayward, CA, 1988.

[19] Persi Diaconis and Laurent Saloff-Coste. Comparison theorems for reversible Markov chains. *Ann. Appl. Probab.*, 3(3):696–730, 1993.

[20] Persi Diaconis and Daniel Stroock. Geometric bounds for eigenvalues of Markov chains. *Ann. Appl. Probab.*, 1(1):36–61, 1991.

[21] Alexander Dinghas. Über eine Klasse superadditiver Mengenfunktionale von Brunn-Minkowski-Lustenikschem typus. *Math. Zeitschr.*, 68:111–125, 1957.

[22] Martin Dyer, Alan Frieze, and Mark Jerrum. On counting independent sets in sparse graphs. In *Proceedings of the 40th Symposium on Foundations of Computer Science (FOCS)*, pages 210–217. IEEE Computer Society Press, 1999.

[23] Martin Dyer, Alan Frieze, and Ravi Kannan. A random polynomial-time algorithm for approximating the volume of convex bodies. *J. Assoc. Comput. Mach.*, 38(1):1–17, 1991.

[24] Martin Dyer and Catherine Greenhill. A more rapidly mixing Markov chain for graph colorings. *Random Structures Algorithms*, 13(3-4):285–317, 1998.

[25] Martin Dyer and Catherine Greenhill. On Markov chains for independent sets. *J. Algorithms*, 35(1):17–49, 2000.

[26] H. G. Eggleston. *Convexity*. Cambridge University Press, New York, 1958.

[27] Tomás Feder and Milena Mihail. Balanced matroids. In *Proceedings of the 24th Annual ACM Symposium on Theory of Computing (STOC)*, pages 26–38. ACM Press, 1992.

[28] Uriel Feige and Carsten Lund. On the hardness of computing the permanent of random matrices. *Comput. Complexity*, 6(2):101–132, 1996/97.

[29] Joan Feigenbaum and Lance Fortnow. Random-self-reducibility of complete sets. *SIAM J. Comput.*, 22(5):994–1005, 1993.

[30] Michael R. Garey and David S. Johnson. *Computers and Intractability: a Guide to the Theory of NP-Completeness.* W. H. Freeman and Co., San Francisco, Calif., 1979.

[31] Leslie Ann Goldberg. Computation in permutation groups: counting and randomly sampling orbits. In *Surveys in combinatorics, 2001 (Sussex)*, pages 109–143. Cambridge Univ. Press, Cambridge, 2001.

[32] Oded Goldreich. *Introduction to Complexity Theory.* Lecture Notes Series of the Electronic Colloquium on Computational Complexity. `http://www.eccc.uni-trier.de/eccc/`, 1999.

[33] G. R. Grimmett and D. R. Stirzaker. *Probability and Random Processes.* The Clarendon Press Oxford University Press, New York, second edition, 1992.

[34] Mark Jerrum. Computational Pólya theory. In *Surveys in combinatorics, 1995 (Stirling)*, pages 103–118. Cambridge Univ. Press, Cambridge, 1995.

[35] Mark Jerrum and Alistair Sinclair. Polynomial-time approximation algorithms for the Ising model. *SIAM J. Comput.*, 22(5):1087–1116, 1993.

[36] Mark Jerrum and Alistair Sinclair. The Markov chain Monte Carlo method: an approach to approximate counting and integration. In Dorit S. Hochbaum, editor, *Approximation Algorithms for NP-hard Problems*, pages 482–520. PWS, 1996.

[37] Mark Jerrum, Alistair Sinclair, and Eric Vigoda. A polynomial-time approximation algorithm for the permanent of a matrix with non-negative entries. *Electronic Colloquium on Computational Complexity*, TR00-079, 2000.

[38] Mark R. Jerrum, Leslie G. Valiant, and Vijay V. Vazirani. Random generation of combinatorial structures from a uniform distribution. *Theoret. Comput. Sci.*, 43(2-3):169–188, 1986.

[39] Ravi Kannan, László Lovász, and Miklós Simonovits. Random walks and an $O^*(n^5)$ volume algorithm for convex bodies. *Random Structures Algorithms*, 11(1):1–50, 1997.

[40] Richard M. Karp, Michael Luby, and Neal Madras. Monte Carlo approximation algorithms for enumeration problems. *J. Algorithms*, 10(3):429–448, 1989.

[41] P. W. Kasteleyn. Graph theory and crystal physics. In Frank Harary, editor, *Graph Theory and Theoretical Physics*, pages 43–110. Academic Press, 1967.

[42] Torgny Lindvall. *Lectures on the coupling method.* John Wiley & Sons Inc., New York, 1992. A Wiley-Interscience Publication.

[43] Torgny Lindvall and L. C. G. Rogers. Coupling of multidimensional diffusions by reflection. *Ann. Probab.*, 14(3):860–872, 1986.

[44] L. Lovász and M. Simonovits. Random walks in a convex body and an improved volume algorithm. *Random Structures Algorithms*, 4(4):359–412, 1993.

[45] László Lovász and M. D. Plummer. *Matching Theory.* North-Holland, 1986.

[46] Michael Luby and Eric Vigoda. Approximately counting up to four (extended abstract). In *Proceedings of the 29th Annual ACM Symposium on Theory of Computing (STOC)*, pages 682–687. ACM Press, 1997.

[47] Michael Luby and Eric Vigoda. Fast convergence of the Glauber dynamics for sampling independent sets. *Random Structures Algorithms*, 15(3-4):229–241, 1999.

[48] Neal Madras and Dana Randall. Markov chain decomposition for convergence rate analysis. To appear in the *Annals of Applied Probability.*

[49] Meena Mahajan and V. Vinay. A combinatorial algorithm for the determinant. In *Proceedings of the 8th Annual ACM-SIAM Symposium on Discrete Algorithms*, pages 730–738. ACM/SIAM, 1997.

[50] Lisa McShine. Random sampling of labeled tournaments. *Electron. J. Combin.*, 7(1):Research Paper 8, 9 pp. (electronic), 2000.

[51] Milena Mihail. Conductance and convergence of Markov chains: a combinatorial treatment of expanders. In *Proceedings of the 30th Symposium on Foundations of Computer Science (FOCS)*, pages 526–531. IEEE Computer Society Press, 1989.

[52] Ben Morris and Alistair Sinclair. Random walks on truncated cubes and sampling 0-1 knapsack solutions. In *Proceedings of the 40th Symposium on Foundations of Computer Science (FOCS)*, pages 230–240. IEEE Computer Society Press, 1999.

[53] J. R. Norris. *Markov chains.* Cambridge University Press, Cambridge, 1997.

[54] Christos H. Papadimitriou. *Computational Complexity.* Addison-Wesley Publishing Company, Reading, MA, 1994.

[55] Neil Robertson, P. D. Seymour, and Robin Thomas. Permanents, Pfaffian orientations, and even directed circuits. *Ann. of Math. (2)*, 150(3):929–975, 1999.

[56] Janos Simon. On the difference between one and many. In *Automata, Languages and Programming (Fourth Colloq., Univ. Turku, Turku, 1977), Lecture Notes in Computer Science, Vol. 52*, pages 480–491. Springer, Berlin, 1977.

[57] Alistair Sinclair. Improved bounds for mixing rates of Markov chains and multicommodity flow. *Combin. Probab. Comput.*, 1(4):351–370, 1992.

[58] Alistair Sinclair. *Algorithms for Random Generation and Counting: a Markov Chain Approach.* Birkhäuser Boston Inc., Boston, MA, 1993.

[59] Seinosuke Toda and Mitsunori Ogiwara. Counting classes are at least as hard as the polynomial-time hierarchy. *SIAM J. Comput.*, 21(2):316–328, 1992.

[60] W. T. Tutte. *Graph Theory*, volume 21 of *Encyclopedia of Mathematics: Combinatorics*. Addison-Wesley, 1984.

[61] L. G. Valiant. Completeness classes in algebra. In *Proceedings of the 11th annual ACM Symposium on Theory of Computing (STOC)*, pages 249–261. ACM, 1979.

[62] L. G. Valiant. The complexity of computing the permanent. *Theoret. Comput. Sci.*, 8(2):189–201, 1979.

[63] L. G. Valiant and V. V. Vazirani. NP is as easy as detecting unique solutions. *Theoret. Comput. Sci.*, 47(1):85–93, 1986.

[64] Leslie G. Valiant. The complexity of enumeration and reliability problems. *SIAM J. Comput.*, 8(3):410–421, 1979.

[65] J. H. van Lint and R. M. Wilson. *A course in combinatorics.* Cambridge University Press, Cambridge, 1992.

[66] Eric Vigoda. Improved bounds for sampling colorings. *J. Math. Phys.*, 41(3):1555–1569, 2000.

Index

GPSR Compliance

The European Union's (EU) General Product Safety Regulation (GPSR) is a set of rules that requires consumer products to be safe and our obligations to ensure this.

If you have any concerns about our products, you can contact us on ProductSafety@springernature.com

In case Publisher is established outside the EU, the EU authorized representative is:

Springer Nature Customer Service Center GmbH
Europaplatz 3
69115 Heidelberg, Germany

Batch number: 10578397

Printed by Printforce, the Netherlands